COMPUTERS IN SPACE

Journeys with NASA

James E. Tomayko

alpha
books

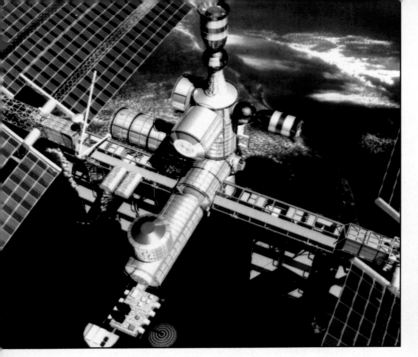

© 1994 by Alpha Books

International Standard Book Number: 1-56761-463-9
Library of Congress Catalog Card Number: 93-74395

97 96 95 94   8 7 6 5 4 3 2 1

Interpretation of the printing code: the rightmost number of the first series of numbers is the year of the book's printing; the rightmost number of the second series of numbers is the number of the book's printing. For example, a printing code of 94-1 shows that the first printing of the book occurred in 1994.

Printed in the United States of America

Publisher *Marie Butler-Knight*

Managing Editor *Elizabeth Keaffaber*

Acquisitions Manager *Barry Pruett*

Product Development Manager *Faithe Wempen*

Manuscript Editor *Barry Childs-Helton*

Cover Designer *Jean Bisesi*

Interior Designer *Barbara Webster*

Layout Designer *Stephanie Gregory*

Production *Gary Adair, Brad Chinn, Kim Cofer, Meshell Dinn, Mark Enochs, Jenny Kucera, Beth Rago, Marc Shecter, Greg Simsic, Kris Simmons, Craig Small, Carol Stamile, Robert Wolf*

Special thanks for photographic illustrations: NASA, Kennedy Space Center, Johnson Space Center, Goddard Space Flight Center, Jet Propulsion Laboratory, Boeing, Marshall Space Flight Center, Langley Research Center, McDonnell Douglas, Microsoft Corporation, US Air Force Museum, IBM Federal Systems Company, Chales Stark Draper Laboratories, Inc., Lockheed Advanced Development Company, Jim and Laura Tomayko, and Paradox Productions.

To my parents,
Inez Tomayko (1915-1991)
and Ed Tomayko,

in gratitude for their support of my interest in science and technology, my education, and for setting an example of integrity, hard work, and faith.

Foreword

Twenty-five years can seem like a very short time, but it's long enough to bring tremendous changes. Apollo XI touched down on the moon about that long ago. The first electronic digital computers, and the first rockets capable of reaching the edge of space, were barely 25 years old at the time.

The Gemini and Apollo programs flew the first manned spacecraft to carry on-board computers. Today, each Space Shuttle carries five. The traditional dialog between pilot and ship is carried on increasingly through computers. Still, as always, space flight means teamwork between human and machine.

This book is about both the Space Age and the Information Age. Practical computers and real space travel took their first steps at about the same time, and each has helped the other grow and advance. Computers have helped us manage the complexities of space flight. The Space Race, in turn, gave the computer industry some tough new problems to solve. The results were two advanced technologies, and a lot of interesting journeys.

I expect the future holds more of them for all of us.

Dr. Buzz Aldrin
Lunar Module Pilot, Apollo XI

Contents

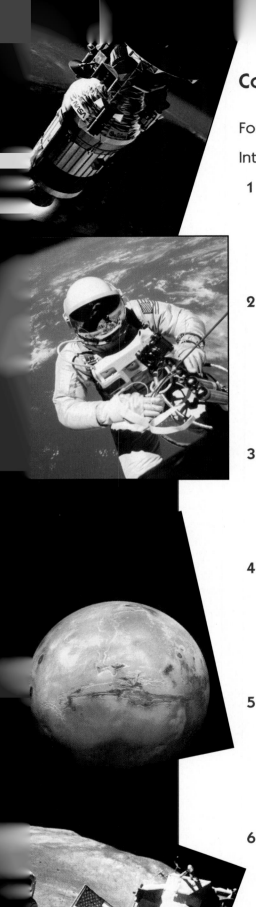

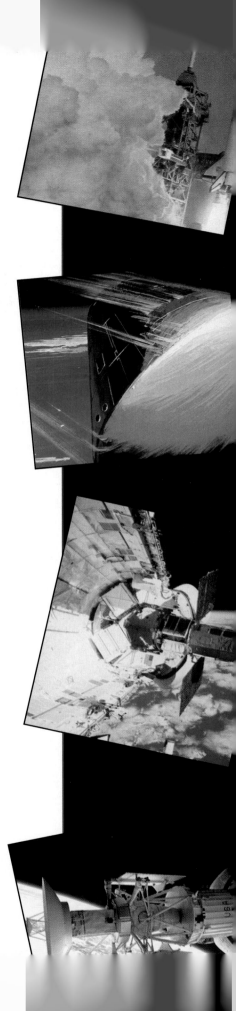

Introduction

"What does a Space Shuttle countdown have in common with tuning my car?"

"Why do modern airliners and the Apollo spacecraft have similar control systems?"

"What component is the same in my auto engine and in a rocket engine?"

The answers to these questions and many more revolve around a machine that has come to dominate late-twentieth-century life: the digital computer. Without the extensive use of computers, a Shuttle flight would not be possible, the exploration of the solar system by robot emissaries could not have happened, and landing on the moon would still be a dream.

In the last minutes before the Space Shuttle takes flight, the carbon-based brains of the ground crew and astronauts defer to the dozens of silicon-based computers that participate in every mission. Too much is happening (and is about to happen) for any team of human beings–however well prepared–to stay adequately on top of events.

Computers in the firing room make the final round of diagnostic tests on the launch vehicle, assisted by the five Data Processing System computers on board the Orbiter (with the help of their many microchip subordinates). Meanwhile, in Houston, the Mission Control computers await their cue: the irrevocable starting of the solid fuel boosters. Just before that spectacular commitment to flight, the computers attached to each Shuttle main engine must agree that the liquid-fueled motors are at full thrust, and that all components are functioning within strict specifications. Otherwise the engines are shut down, and a disappointed group of astronauts must wait for another day to enjoy their ride to space.

Yet, despite the usefulness of computers in space flight, the machines the astronauts "fly" are actually less powerful–and much smaller in terms of memory capacity–than many of the desktop PCs that now populate the Earth in their millions. NASA has avoided going beyond the state of the art wherever possible–in part not to endanger crew safety or science objectives, and in part to save costs. Nevertheless, NASA's needs led an accelerated, more widespread adoption of many technologies related to computing. These include the microchip, software engineering, distributed systems, and the concept of active flight control. These technologies touch our lives every day.

This book shows how computers are used in all phases of space flight–both manned and unmanned–by following the 1989 launch of the *Galileo* probe to Jupiter by Space Shuttle mission STS-34. The five Shuttle astronauts of STS-34 were Don Williams (commander), Mike McCulley (pilot), and three mission specialists: Ellen Baker, Franklin Chang-Diaz, and Shannon Lucid. After sending the probe on its long journey, they returned to Earth. *Galileo*, the most sophisticated unmanned spacecraft ever built, still en route, is traveling to the gas-giant planet for extensive explorations of its moons, and to deploy a probe into the Jovian atmosphere.

As we see the preparations for flight by both astronauts and launch-operations engineers, the teams of controllers working with the manned and unmanned spacecraft, and the actions of crew (both human and robotic), the true nature of computing in space flight becomes evident: it is a pervasive, indispensable technology. No less important is the impact NASA's use of computers has had on the history of computing in general–another theme we will explore as we follow *Galileo's* trail to the depths of the Solar System.

The crew of STS-34 that launched the Galileo probe to Jupiter: Donald E. Williams, Mission Commander; Michael J. McCulley, Mission Pilot; Mission Specialists Shannon W. Lucid, Franklin R. Chang-Diaz, and Ellen S. Baker.

Acknowledgments

I unearthed (bad pun intended) much of the information in this book as part of a contract I had with NASA from 1983–1985. Ed Ezell, then head of the history office at Johnson Space Center (and more recently at the Smithsonian Institution), was especially encouraging. I'm pretty sure he convinced Monte Wright, the chief NASA historian at the time, to give me the award. The resulting work is published as a contractor report for distribution internal to NASA and its contractors, and also as Volume 18, Supplement 3 of the *Encyclopedia of Computer Science and Technology* (Allen Kent and James G. Williams, eds., Marcel Dekker, Inc.). Readers interested in detailed source references (and the ins and outs of some technical decisions) can look there. They can also find acknowledged the legion of NASA and contractor technical personnel who helped me in my travels to nearly all the agency's centers and archives, as well as my valued research assistants. Most of the original source material from the NASA contract is in the Special Collections of the Ablah Library, The Wichita State University, Wichita, Kansas.

Since the end of the NASA contract, I have had help both in finding new material and in staying current. Debbie Martin, a research assistant and now elementary school teacher, located and organized sources on fly-by-wire technology. Jim Morris and Gavin Jenney, both of whom worked in the Flight Dynamics Laboratory at Wright-Patterson Air Force Base, provided both history and technical advice on that subject. Tony Macina and Jeanie Clifton of IBM in Houston kept me up-to-date on happenings with the Shuttle on-board software. Dr. Helmut Hoelzer,

former head of computation at the Marshall Space Center, has been a continuing source of information and insight. Bob Floyd of Johnson Space Center helped me get back up to speed on the Mission Control computers and historical experiences with the Shuttle flight-control computers.

The illustrations in this book are largely from NASA archives or from NASA contractors. We all would like to thank all of the public-relations folks who so generously donated time and resources.

Skip Shelly, a graphic-design genius at the Software Engineering Institute, donated valuable time and advice in helping me get some difficult illustrations ready for scanning.

I would especially like to acknowledge David Craig, whose infectious interest in space-flight computing and lively correspondence kept me from wandering away to other pursuits. He has gathered a terrific collection of material on the Apollo spacecraft computer software, among other things.

Working with Alpha Books means working with a team. Their names appear elsewhere, but I would like to single out manuscript editor Barry Childs-Helton (who suggested this subject to the publisher, and made me humble in the face of his smooth translations of my technical prose), Faithe Wempen, the product development manager (who was the "voice of the reader"), and Barry Pruett, the acquisition editor, who doggedly pursued the hundreds of illustrations I specified—a task I am not sure is in his job description, but just page through the book and see how well he did it.

Lastly, I would like to thank my wife, Laura, who kept up my spirits in those long winter nights in front of the merciless monitor of my computer.

Basic Virtual Reality: Computers Simulating Space Flight

1

The astronaut crew feels the heaviness in their bodies as the Space Shuttle accelerates after liftoff. Out the forward windows, Earth is "above" and space "below"; the Shuttle is flying on its back. After the last rocket engines shut off, the astronauts feel light again. Too light for Earth's gravity. They are in space, the forward windows dark with its blackness. The brilliant white cargo bay doors can be seen opening in the rear windows.

Suddenly, the doors stop moving. They're stuck. The crew responds to the alarm, and tries to troubleshoot as quickly as possible: half open, the radiators in the door lining cannot keep the Shuttle cool. Half open, the orbiter cannot return to Earth. Either way, it's big trouble. The astronauts are just about to reset some circuit breakers when *the side hatch opens*. Their four hours are up, and it's time for the next crew to be tortured by trainers in the Shuttle Mission Simulator, motion-base version.

The real thing requires realistic training. The Space Shuttle astronauts spend many hundreds of hours in simulator training before their actual flight.

Simulators are where the astronauts do their "real" work on Earth. For every mission, it's hundreds of hours of practice. The Apollo moon landing program probably had the highest ratio of "sim time" to flight time, but the Shuttle training is by no means less taxing overall.

Only the use of computers—in big interconnected networks—can create a realistic simulation like the one just described. For nearly every activity that has a significant element of danger (or is simply too costly to practice on actual equipment), there is now a simulator.

Every rocket designed since the V-2 has flown thousands of times inside a computer. Every pilot moving into a new high-performance aircraft runs up hours and hours of imitation takeoffs and landings. Nuclear reactor operators train for the worst. It's a lot cheaper, and you don't bend the sheet metal or kill a few thousand people every time you make a mistake.

Not all simulators are like the *motion-base*, high-fidelity Shuttle Mission Simulator. That one has a complete orbiter cockpit, and is mounted on a tilt table to let it imitate the spacecraft's movements. There are also *part-task* trainers that break down activities into components; the trainee can practice these segments over and over, until ready to combine them in a full simulation. *Fixed-base* simulators (complete except for the capability of tilting the flight deck) provide further practice.

Readers who own Microsoft Flight Simulator, or have seen it in action, know generally what a fixed-base simulator can do. It can involve you pretty deeply in the game, but the floor stays straight and level. In the motion-base simulators, the deck is inclined to put pressure on the crew's backs to simulate acceleration, and later inclined forward to give them a constant feeling of falling akin to weightlessness. There are also detailed engineering simulators to help integrate improvements in avionics and hardware of all types.

The motion-base version of the Shuttle Mission Simulator. The cockpit is the white box in the background. Mission controllers sit at the various consoles in the foreground.

Microsoft Flight Simulator, a PC program with high-resolution graphics for practicing flying many different aircraft.

Most pilots can get comfortable with flying a low-performance airplane like a Cessna 152 or a Piper Tomahawk without the aid of a simulator. As long as the instructor is sitting in the right seat, nothing too bad happens. Each higher rung on the performance scale is marked by further instruction, up to the point where a fledgling pilot is checked out for "high-performance, complex aircraft," or those with a minimum of 200-horsepower engines, retractable landing gear, and constant-speed propellers. That is the most complicated level of pilot proficiency the Federal Aviation Administration demands for single-engine aircraft.

It is, however, a long jump from there to thousand-horse-plus behemoths like the P-47 Thunderbolt or P-51 Mustang fighters of World War II. It's a similar jump from twin-engine turboprop business aircraft to a twin-engine heavy like the Boeing 767 airliner; a pilot—no matter how talented—stretches his or her ability. Simulators bridge the gap. Young first officers can sit in the right seat of a 767 simulator and practice approaches all they want. No costly fuel to burn, no traffic (unless the instructor throws some into the sim), and no penalty for digging holes on the end of the runway. As a bonus, "flight time" in a high-fidelity simulator counts as actual hours in the pilot's logbook. That's how good the machines can be.

Aircraft simulators take pilots where they have been and will go again. Flying an approach to Pittsburgh in a 767 simulator is especially good

Avionics

As aircraft became more and more dependent on electronic devices for navigation and other functions, the word "avionics" came into use. It is short for "aviation electronics," and is used as a general term for navigation radio receivers, radios in general, global-positioning-system receivers, and electronic flight-control systems.

Computers in Flight Training Simulation

Without all these different ways of practicing, astronauts could get into deep trouble on actual missions—really quickly. The instruction given to them follows the historically-successful techniques of flight training. The idea is to always "stay ahead of the airplane." To be prepared and stay in control, the pilot must know what is likely to happen in the next few moments.

training if you have flown the same approach in a Jetstream a few times "for real." You can readily see the differences and account for them.

The first spacecraft simulators, however, had to take astronauts where they had *never* been before. The lunar landing simulator was probably the single most important training device of the Apollo program. Errors made while landing on the moon would waste fuel and air, and were likely to be irreversible. Lunar landing was also the activity with the fewest earthly counterparts. Vertical landings using reaction engines like jets and rockets had a history that would fit on one page. One or two vertical-takeoff-and-landing airplanes had been built in the 1950s, but no astronaut had flown them. Without the simulator, there would have been no way to prepare to "stay ahead of the lunar lander."

As aircraft and spacecraft became more sophisticated, a pilot had to handle an increasing workload safely in a complex, quickly-changing environment. Simulators for aircraft had become common in the 1940s, due to the wartime demand for pilots. The "Link Trainer" was the first popular flight simulator. This type of trainer—an enclosed box with real instruments driven by fixed electronics—operated using *analog circuits*.

Analogs are imitations of reality. In an electronic analog, the trainee moves a control stick embedded in special transformers, generating a voltage proportional to the movement. That voltage is routed through a hard-wired electronic circuit, which continuously calculates what effect the movement would have on specific control surfaces and instruments. Instruments display the effect to the trainee, or the simulator moves.

The beauty of these analog devices is that they communicate with one another using the same

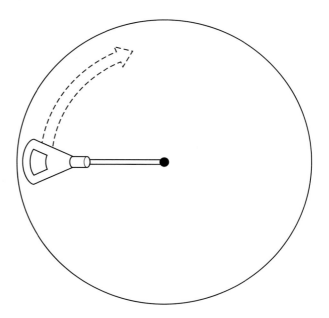

The Mercury cockpit could be mounted at the end of an arm that spun around a central axis, creating simulated acceleration forces.

"language," a voltage level. Therefore the circuitry, once made and "balanced" to reflect reality, is set for the duration. The bad news is that any change in aircraft handling or instrumentation means redesigning and rewiring the simulators, perhaps extensively. Another downside was size. An analog version of the F-4 fighter-bomber simulator grew to a small houseful of vacuum tubes and circuits—rivaling the bulky ENIAC, the first all-electronic *digital computer* built in the mid-1940s.

Other affinities were growing between simulators and computers. The Navy funded Project Whirlwind at MIT in the late 1940s and early 50s, seeking to use digital computers as the basic means of doing the calculations that created what we might call the simulation's "virtual reality." Critical breakthroughs in computer technology resulted, especially in memory. Whirlwind also led

Shuttle Mission Simulator, fixed-base version. The two boxes that look outlined in black (upper center) are window-view generators, and part of the computer hardware is in the racks to the left of the controller stations.

simulator designers to think of new ways to incorporate computers. Even so, by the time NASA needed trainers for Project Mercury (its first manned flight program), analog systems still dominated simulation.

Analog Versus Digital

It may be easier to think of *analog data* as "continuous," and *digital data* as "distinct." Any analog device receives and displays information in an uninterrupted way. For instance, an analog clock has a second hand that sweeps around the dial in a smooth motion. It is very difficult to tell when one second ends and the next begins. A digital watch, on the other hand (pun intended), shows clearly the beginning of each second as the numbers simply change.

Space medicine was a new field in the late 1950s, long on speculation and short on real data. Doctors worried about the effects of acceleration and weightlessness. In space flight, acceleration is short and weightlessness long. On Earth, the opposite conditions prevailed. You could build a big centrifuge and crush your astronauts for hours on end, but weightlessness could only be achieved for a few seconds by putting an airplane on a roller-coaster trajectory.

Since the most critical mission phases in a simple space flight program like Mercury are powered flight and then re-entry, the Mercury simulators concentrated on those. A mockup of the space capsule mounted on a centrifuge gave the pilot a feel for how it would be to flip switches or pull the abort handle when his arm weighed a hundred pounds or so. Other specialized trainers drilled prospective astronauts on re-entry procedures. Two fixed-base trainers, containing analog circuits, let them practice procedures for an entire flight. One was at Langley Space Center in Virginia, another at Cape Canaveral for last-minute cramming.

Training for Gemini

NASA's second manned flight program, the Gemini series, was effectively a bridge to Apollo; it provided practice in critical maneuvers needed to fly to the moon (see Chapter 5). Gemini simulators were themselves a bridge, between the all-analog world of Mercury and the nearly all-digital world of Apollo. Now there would be two pilots in the spacecraft, and it could actually change its orbit, something the Mercury spacecraft could not do.

As it did for Mercury, NASA built a pair of *procedures simulators* for Gemini—installing one in its new Manned Space Flight Center (now Johnson Space Center) in Houston, the other at the Cape. Since one major objective of Gemini was to join up with another spacecraft, the space agency also needed motion-base simulators; practice was essential for the crucial tasks of rendezvous and docking. Another motion-base simulator would specialize in those two heavily-practiced parts of the flight, ascent and entry. Singer Corporation's Link Division had the contract to build the key components for all of these.

Digital computers had three main tasks in Gemini simulators: imitate the on-board computer, control the views from the windows, and handle communications data from the outside world. For the first two tasks, NASA bought DDP-224s built by Computer Control Corporation (later absorbed by Honeywell). These were among the first machines built for use by engineers applying control theory, instead of for data processors doing payroll programs. Digital Equipment Corporation was another supplier of this type of computer, starting with the PDP-1. Small in comparison to the mainframes of the day, these *minicomputers* rapidly became useful in manufacturing and design; even today's "workstation" PCs have not totally replaced them.

The DDP-224s were quite capable calculators. They had a 24-bit word length and could do floating-point arithmetic. That made them flexible

An outdoor test of the lunar rover vehicle.

Giant Monuments on the Moon

In the mid-1970s, the Apollo-Soyuz Test Program was to link up American and Russian spacecraft in orbit. Soviet cosmonauts visited various NASA centers on a goodwill tour surrounding the upcoming flight. One cosmonaut gave his insignia pin to a U.S. astronaut who was showing him around the Marshall Space Center. This cosmonaut later got a chance to "drive" the lunar buggy used in Apollo in a simulator that moved cameras over fake moonscape for the forward views. The cosmonaut drove around a boulder and suddenly slammed on the brakes. In front of him was his cosmonaut insignia pin, now standing many stories tall on the moon! The American astronaut had put it on the terrain model as a joke.

and powerful, two of the most important attributes a simulator computer must have. The Geminis were the first manned spacecraft to carry an on-board computer for navigation and control; the DDP-224s' first task was to imitate it.

Instead of having the actual on-board computer in the trainers, the designers "functionally simulated" it. This means the real computer's algorithms and programs were adapted to the DDP-224, and executed just as though the real computer were doing the work. Fortunately, the relatively simple Gemini computer could be easily "faked" by the simulator machine. One thing was immediately clear: it was easier to imitate a digital computer with another digital computer than with an analog machine.

The DDP-224s' second major task was to control the "scenery" outside the astronauts' windows. Even though the Gemini spacecraft had just two small viewports, generating views for them digitally was not an option in the early 1960s. Instead, television cameras in another room captured images of models and paintings: other spacecraft, the Earth, and the sky. The cameras were mounted on moving tracks, so they could follow the Earth or show the different views of another object from the point of view of the spacecraft as the spacecraft "moved." This is how all aircraft simulators worked at the time: a camera moved over a terrain model, mimicking the motion of the airplane. It could be surprisingly realistic. After "flying" along for awhile, watching the scenery change below, pilots would lose themselves in the charade—and grab their coffee cups when the airplane "banked!" In the Gemini simulator, the computers controlled the cameras' movements as needed, giving the astronauts a "realistic" view of the outside world.

To give the astronaut crews and their ground controllers practice working with one another, Gemini simulators required that they be able to communicate with Houston and all Earth-based tracking stations in the network. A special computer had to be built just to handle data communications. With the three DDP-224s already in the simulator, this cluster of computing power was one of the largest ever built for NASA. Though Mission Control computers would soon outstrip it in memory capacity and processing speed, this complex system still had few commercial antecedents.

The Lunar Excursion Module Simulator flies a night mission. This multiple-exposure photograph shows the simulator being lifted up into the structure that balanced out enough of Earth's gravity to simulate the handling of the Module during descent.

This new approach gave astronauts at least two terrific training advantages: they could sit in a simulator that looked and handled like the real thing, and they could work with all the engineers and technicians involved in the flight. It did not take long to roll all the lessons learned from Gemini into the Apollo mission simulators.

If This Is Tuesday, I Must Be in Houston

Flying to the moon and back proved vastly more complicated than science fiction could have predicted. Apollo-era science and technology was more than adequate to accomplish the mission, but not to make it easy. The crews had to memorize huge amounts of data and long lists of procedures. Numbers could be in different bases, scales, or types—and there was no way for the on-board computer displays to indicate which they were. You had to know from context.

In addition, the formidably busy schedule of flight activities demanded great piloting skills (see Chapter 6). NASA recognized early on that the only way to gain proficiency was training, training, and more training. This meant simulators, simulators, and more simulators: *15* in all. Even some of the simulators needed simulators!

A bizarre-looking four-legged creature of struts and landing pads, with a jet engine mounted vertically in the middle, could be used to practice landings with five-sixths of Earth's gravity balanced

The Lunar Excursion Module Simulator's supporting structure at the Langley Research Center. Note the imitation lunar surface under the beams. This simulator is now a National Historic Landmark.

out. That way the pilots could get used to a descent in the moon's one-sixth gravity. But it was so difficult to fly, even for skilled pilots, it needed its own simulator. Neil Armstrong had to eject when the "Flying Bedstead" went out of control; the first man to set foot on the moon was nearly killed by his dangerous (but necessary) training.

Fully 80 percent of Apollo training time was in one sim or another. The Apollo simulators required much more computing power than the Gemini program. First, there were two spacecraft in each lunar mission: Command Module and Lunar Lander. Instead of just three DDP-224s and the communications computer, each Command Module simulator needed *five*. Three of them simulated spacecraft systems, one imitated the on-

board computer, and one represented the launch vehicle. The Lander simulator needed two for the spacecraft systems, and one for its on-board computer.

When a full simulation of a lunar landing was in progress, with Mission Control on-line as well, *ten* computers had to interact. Developing the software to drive these computers turned out to be a significant part of the program. There were 175 software engineers and 200 hardware engineers at work on the simulator designs. The result was 350,000 words of software—all in *assembly language!*—tiny by more modern standards, but (by the same standards) obscure and difficult to troubleshoot.

One of the biggest problems in building the simulator was imitating the Apollo on-board computer. The first try was a functional simulation like the one in the Gemini simulators, but that turned out to be too hard to do well. The Gemini computer had been slow and not too complicated; a simulation that merely faked its functions performed reasonably in both complexity and speed. In contrast, the Apollo computer had a lot more software, and ran in a relatively fast machine. A purely functional simulator would have been too slow *and* complicated to provide effective training.

An Imitation for the Imitation

Maybe it would have been better (and simpler) to put some real Apollo computers into the loop. But the cost and difficulty of maintaining software for such a radically different computer sobered the Singer-Link folks who had to put it all together. It was already taking four months to write code that functionally simulated the flight software load for one mission.

Finally, W. B. Goeckler (in the Apollo program's Systems Engineering office) had the idea of mak-

ing the DDP-224 computer "think" it was an Apollo computer, and simply having it execute the Apollo code. A new engineer named Jim Raney got the task of seeing if this could be done.

Bits, Words, and Floating-Point: Data and How It Is Manipulated

The computing field has its own meanings for both familiar and unfamiliar terms. A *bit* is short for "binary digit," which is a one or a zero. Those are the only two numbers a computer uses for calculations. A *word* is a single unit of information in a computer. Small microprocessors, like those in the first PCs of the early 1980s, often had 8-bit words. The bigger the word, the more accurate the calculating ability. Any machine's calculating range can be increased by using *floating-point arithmetic*. This method is a lot like scientific notation; when storing a number, the computer stores the location of its decimal point, and the values of its significant figures. Since the decimal point can move left or right as necessary, this type of data storage is called *floating-point*.

He had to change some things on the DDP-224 to do it. The Apollo machine was not a floating-point machine, so the Honeywell computer could not use its float capability. Also, the Apollo computer used a 16-bit word (one parity

The Apollo Mission Simulator cockpit for the Command Module is in the upper right of this photograph. Note the stream of thick cables cascading out of the support structure in the upper left.

bit, one sign bit, and 14 instruction or data bits). The 24 bits of a DDP-224 had to be converted to this format: a bit that signaled it was an Apollo word, nine identical sign bits to fill up the extra space, and 14 bits of Apollo code in the last 14 bits of the word.

The fixed Apollo flight program went into the upper half of the DDP-224's 64K memory; all the erasable data needed was in the lower half. An 8K block of that half was used as common memory among the many simulation computers, so that they could share information without duplicating it. It was a simulation within a simulation; few thought it would work. In the end, the simulated computer actually ran *faster* than the real computer. Crews complained that when they got into the real spacecraft, the computer was annoyingly slow. The solution was to slow down the simulation, and get them annoyed *before* liftoff!

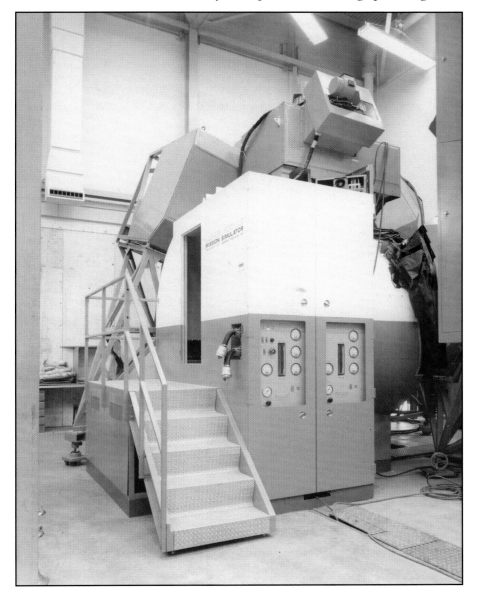

The Apollo Mission Simulator cockpit for the Lunar Module.

"Is It Live, or Is It Memorex?"

NASA had a problem when it tried to build a simulator for the Space Shuttle. Nobody wanted to do it. No surprise; the author once had to box and move the original requirements and specifications. It took a full day of diligent packing, sealing, and marking. If the simulator were as large as it was complex, any contractor was doomed.

Singer finally overcame its shyness, and bid. They came back with a spec for a computer network that is daunting even today: four UNIVAC 1100/40 mainframe computers, 15 Perkin-Elmer 8/32 minicomputers, and five IBM AP-101 on-board computers for the fixed-base version of the simulator (which seats four, and has full window views front, back, and up). The motion-base simulator was the same, though it used fewer minicomputers: since only two seats and the forward

Interior of the Lunar Module simulator.

windows were in the sim, it did not need to create as many graphics (in this case, "views").

None of these computers are built today (even the on-board computers have been upgraded). In fact, neither Sperry nor Perkin-Elmer is still in the same *sort* of computer business as when they built this system. To maintain the operating system (which had been dropped as a product) and keep the simulator computers running, NASA had to hire Sperry engineers away from the UNIVAC shop. (We'll see later that this is by no means an unusual move for the space agency.)

NASA's basic problem with the simulator computers was that they kept wearing out. It did not take long to figure out that a functional simulation of the Shuttle's AP-101 computers would not work.

The simulated AP-101s were just as fast as the UNIVACs, which meant a less accurate simulation. Singer designed the real computers into the system. They also tried using the real magnetic tape mass memory used on the Shuttle (see Chapters 7 and 8). The tape drive, designed for only a few minutes of use each flight, crumpled under the 16-hour-a-day pounding of training. It had to be replaced with a special minicomputer that held the current software load, and was slowed artificially to match the speed of the tape.

The actual AP-101s did not fare much better. It is a struggle to keep five computers up-and-running in each of the big simulators. If any are available beyond the ten needed in the fixed- and motion-base simulators, they are used in the "Spare Parts

Simulator" (a facetious name for a *task trainer* that helps astronauts practice guidance and navigation procedures).

Even Smaller Computers Offer Help

As PCs became widely available in the early 1980s, NASA tried to incorporate them into training. A lot cheaper than the big mission simulators, they were also more capable in certain limited areas of training. Copying ideas from the PLATO systems used for education and training, NASA used PCs with good graphics and touch-screens to train beginning crews on specific tasks. PCs were able to put a screen from the on-board computer system on the same display with a graphic of a meter or switch, so the astronaut could see how they went together.

When the Shuttle simulators came on-line in the late 1970s, they led the world in technology. With the improvement in computers and graphics throughout the 1980s, they are now in the "middle" category. Aircraft simulators for transports and military airplanes have better graphics and responsiveness. Even Microsoft Flight Simulator uses fractals to depict terrain more accurately. (At least TV cameras scooting over terrain models are a thing of the past.) Someday soon, the simulator that used to be the center of a network of big mainframes will be a single processor, running a display inside a pair of goggles and reacting to a control stick. Crew members will see it all in front of their eyes: cockpit, outside views, their hands on switches.

Growing the Space Shuttle: Engineering Simulators

Modeling is one of the most important design techniques in architecture and engineering. Few architects construct buildings without a scale three-dimensional model to help client and designer visualize the real

thing. Even computer-assisted designs are translated into physical models at some point. (Since architects can only erase their mistakes with jackhammers, this is a money- and time-saving approach.)

The Space Shuttle is like a building in that way: it had to work the first time. Though a building contractor can always move a wall and refurbish a bathroom later, the Shuttle program had to achieve something no other manned spacecraft has ever done—fly with a human crew on its first mission—and it succeeded. (Even the Soviet shuttle *Buran*—though it is nearly an exact copy of the U.S. design and technology—was unmanned on its first flight.)

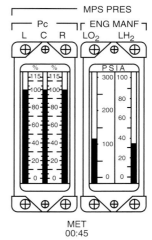

The manifold pressure gauges (top) and ascent trajectory screen (bottom) appear together on the simulator to help trainees learn their relationship.

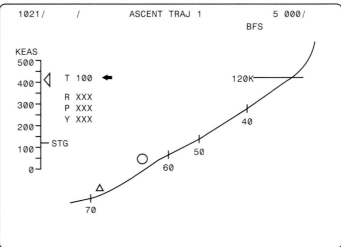

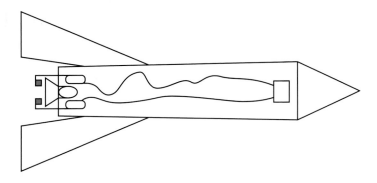

The vanes in the V-2 rocket exhaust steered the missile on the basis of commands from its analog computer.

Getting NASA to be so confident in the Shuttle design was possible because it was "grown" over time, in engineering simulators driven by computers that made spacecraft systems "believe" they were in flight. STS-1, the first mission of the new Space Transportation System (the Shuttle program's official name), was the crowning achievement of such technology, which had its start 40 years earlier in Germany.

The Birth of Rocket Simulation

By the late 1930s, the German army was building the ultimate long-range artillery: a ballistic missile called the A-4. Later renamed the V-2 by Hitler, the rocket was the first vehicle—weapon or not—to touch the edge of space. Wernher von Braun and his team of engineers were close to success in completing the missile design.

The rocket engine worked most of the time. The real problem was in stability and guidance. In the last century, Congreve artillery rockets (like those that gave "the rockets' red glare" to the U.S. national anthem) had used long sticks on their tails to balance in flight. Later rockets used fins, and the V-2 had them. However, a rocket has to be *moving* through the air before its fins will help steady it. That first few seconds of flight, as the engine comes up to full thrust and the craft accelerates, is a critical time of instability.

The Germans developed mechanical controls. Von Braun's team put graphite vanes in the rocket exhaust to act like rudders on a boat, and keep the rocket balanced like a broomstick on someone's finger. These vanes connected to a mechanical system that sensed motion and gave feedback to the vane actuators. The problem was it did not work quickly enough (or well enough) for deployment to the field. It took a long time and great care to align the vanes before launch, and the system was tricky. (Later the Allies would witness, on captured German films, many V-2s rising two or three inches, then falling on their sides and exploding.)

Just as the control problem reached its peak, a recently-arrived young engineer named Helmut Hoelzer tried a different approach. He designed an analog circuit that used electronic components to *model* the control system and the rocket's motion. Over time, this simulator grew into a flight control system for the V-2.

Twenty years later, Hoelzer would be head of the Computation Laboratory of the Marshall Space Flight Center in Huntsville. There he bought one of the first IBM 7090 mainframe computers to perform engineering simulations of the Saturn class of launch vehicles. His analog computers had served well in designing the Redstone and Jupiter rockets, but the Saturn was something else. The Saturn I had eight engines in the first stage, each one roughly as powerful as any previous engine built at Marshall. It was a challenge to existing computer technology.

In an analog computer simulator, the components have to be rearranged (and maybe redesigned) to reflect engineering changes. This process is difficult, and sometimes lengthy; in effect, you are redesigning the computer as well. In a digital computer like the ones now in our homes, changing the computer's ability to do something means simply changing *software*, a much more tractable medium than circuit boards and wire. In designing what he knew to be a complex machine that would stretch the limit of technology, Hoelzer wanted to exploit the flexibility of software. Even though none of his computers had even a fifth the memory of the smallest PC today, he succeeded. There are no films of Saturns falling off the launch pad. All Saturn I (and later Saturn V) rocket launches were successful. This is truly an incredible record, and it helped assure computers a place as essential tools in space exploration.

The Model Orbiter

Using engineering simulators to design manned spacecraft usually starts with the cockpit. That is where the pilots have to work; it's the part they criticize most! The first Shuttle cockpit simulator began operating in 1968, well before engineers settled on the final configuration of boosters and orbiter. It took a Control Data Corporation Cyber 74 mainframe and four minicomputers to drive the simulator and its displays. By 1972, the overall design of the orbiter was set. As the required type and number of components became available, NASA and Rockwell International (the prime contractor for the Shuttle) would install them in what became the Shuttle Avionics Integration Laboratory, or SAIL.

The SAIL is one of the biggest engineering simulators ever built. It contains every electronic component on the spacecraft. If you visit the SAIL, you can see the rough outline of the shape of the spacecraft. The on-board computers, and most parts, are exactly where they are on the real

Dr. Helmut Hoelzer scans printouts of astrodynamics simulations at the operator's station of the IBM 7090. Dr. Rudolf Hoelker is at the blackboard. A model of the Saturn I is partially visible to the right.

Astronaut John G. Creighton in the Shuttle Avionics Integration Laboratory cockpit.

orbiters. Other computers have to imitate some of the components that interact with the on-board machines (such as the reaction control jets that change the orbiter's attitude in flight). Analog-type computers often represented these imitated components at first, but digital machines have replaced most of them. These devices must be placed so that the signal delay is exactly the same as on the real spacecraft. This means that some are yards away from the orbiter outline in the SAIL.

For the first decade or so of the Shuttle program, the SAIL helped verify the design of the electronics on the Shuttle. Since it is a complete simulation, it can even run flight software and be used for crew training. Many of the early software errors were found in the simulator. The first color window displays were in the SAIL. These displays are also very high-quality, so crews training to use the Shuttle's remote-controlled arm preferred to do so in the SAIL.

In its present form, the simulator can be used to try out any change in an electronic component; the change can be validated before it is used in actual flight. The SAIL is also the only simulator that allows the engineers who do the Shuttle's preflight checkout to test the avionics programs fully (see Chapter 3).

When STS-34 lifted off with *Galileo* in the cargo hold, the crew of five probably had a feeling of déjà vu. Hadn't they been there before? When an alarm went off on the fourth morning of the flight, warning about the gas generator, did the crew think (as they awoke) that it was yet another sadistic trainer's test? If so, it never hampered their performance. They reacted quickly and well to all malfunctions and alarms, and launched *Galileo* to Jupiter on schedule. They acted like every trainer's "dream team."

The real Saturn I just before ignition.

Ready for Liftoff! Using Computers to "Preflight" Giant Rockets

2

If you are like most people, you hardly glance at your car when you hop into it, start the engine, and go. You know that the worst systems failure likely to happen is a breakdown, causing you to drift slowly to the side of the road and have an inconvenient delay. And even *that* does not happen often.

If you have flown, you might have noticed that a pilot takes a little more care. He or she walks around the aircraft, looking at all the control surfaces, the tires, and the brakes. Back in the cockpit, the pilot sets switches and instruments, following a checklist to make sure everything is working and ready to go. If it is a piston-engined airplane, the pilot does a "runup," taking the engine to a higher RPM setting and checking the magnetos. All this is done to make sure anything broken can be detected and fixed before pilot, crew, and a couple of hundred passengers are at 30,000 feet where mechanics are scarce.

The Apollo-Saturn V spacecraft at the moment of launch.

Even so, some broken things cannot be found solely by inspection. Once, all the electric instruments failed in flight on a piston airplane I was piloting. A chunk of interior wall had separated, shorting out the alternator. (The prop kept turning; once started, the engine is fired by magnetos.) I had to extend the landing gear by unlocking it and stalling a few times to shake it down. Visual inspection could not detect this problem; I could not have known it would happen.

Special test devices are needed in a more complex system. Auto manufacturers know the average driver consults no checklist before driving, so they build the checklist in (seat belt lights, for instance). With the new computer-controlled ignition and air/fuel systems, they can put self-tests into the hardware. That "service engine soon" light means the self-tests have turned up something to worry about. Airplanes also have built-in test equipment, most often driven by software, to check components invisible to the pilots. In military airplanes like the B-1 bomber, the test equipment runs during the entire mission, and provides the maintenance personnel a list of all the "squawks."

Complex as they are, autos and airplanes (even the most modern) seem simple compared to the Saturn V rocket that sent astronauts to the moon. Preflighting—preparing it for launch—was vastly complicated. There were hundreds of thousands of parts; rocket engineers cared about every single one. It was not enough to check entire components such as rocket engines; components-inside-components (valves, fuel lines, relays) also had to be checked—and quickly, while conditions were right for launch.

The importance of computers to rocketry and space flight cannot be overstated—especially in the crucial area of preflight checkout. Wernher von Braun himself said that NASA owed the Saturn V's success to "our automatic checkout procedures"—and without significant computer power, these would have been impossible to implement.

Manual Versus Automated

There are two types of checkout done on a space vehicle. The first is at the factory. If anything is broken, that is the best place to find

Liftoff of Apollo XI on the way to the first landing on the moon.

it, since spare parts and skilled technicians are right there to set things right. The second checkout is at the launch site, during the "countdown" to the zero moment when the engines start. This checkout ensures that the process of shipping the rocket stages to the Cape and stacking them has not damaged them.

Such a division of labor had been necessary in rocketry since the 1940s. In the last chapter we saw how an analog computer simulated the V-2's motion for the Germans, helping them improve

the rocket's guidance system. Once built, the guidance circuits could be tested using the same type of computer. Technicians could hook up the analog computer simulating the rocket to the analog computer on board, and observe the results. During checkout, an engineer would then compare actual voltage readings to the expected readings. Anomalies were dealt with by repair or replacement.

Exterior view of the Project Mercury blockhouse.

Sometimes a Kludge Does the Job

Whether computer or missile, a *kludge* is a device built of mismatched parts. Some of the computers used in aerospace are kludges because they have components drawn from a variety of analog and digital devices, connected in a way that is sometimes awkward. The Jupiter-C rocket that orbited Explorer I was a kludge made up of a Redstone booster, Sergeant solid-rocket motors in various configurations for upper stages, and a name "borrowed" from another missile in order to hide the source of funding.

Even so, the kludge brought respectable results. Launched in January 1958, the thirty-pound Explorer I satellite was the first U.S. spacecraft to reach orbit.

Readying the Explorer I satellite for launch at Cape Canaveral, 1958.

Cables and More Cables

As rockets grew more complex, the number of test points in a vehicle grew rapidly, and the engineers' expertise increased. Throughout the 1950s, each preflight checkout point had its own discrete line, running from the rocket to the blockhouse where the engineers sat at their test consoles. Countdowns became highly choreographed ballets, each engineer doing his part, coming together at the defining moment of liftoff.

Vanguard, the first U.S. artificial satellite launch program (1957), had about 600 test points in its three-stage rocket and tiny, 3.5-pound satellite. A few dozen engineers could check out the vehicle in a few days of countdown. Although beaten to space by the Soviet Sputnik launch (and by the highly successful kludge the von Braun team put together in the ensuing "missile gap" panic), Vanguard still represented the peak of technology for its time.

By the time the Apollo spacecraft came into use, there were 2,500 test points in the manned section alone. In the Saturn V booster, there were 5,000. If done manually, adequate preparation of the rocket would have taken a year! Even so, there was a huge dispute over the efficacy of automated testing. Some engineers felt that computers could not replace expert judgment. Others argued that manual checkout was so repetitive and tedious that technicians often missed failed parts anyway.

In the early 1960s, everything was still manual. A Saturn I booster had hundreds of lines running to it from the blockhouse. A Mercury spacecraft during testing had so many cables snaking out from the hatch that an astronaut or technician could barely squeeze inside to help with the tests. Something had to be done. Finally, a design decision at the launch site itself settled the issue.

Part of the launch processing facilities at Kennedy Space Center, built for the Apollo program and now used for the Shuttle. The Vehicle Assembly Building has the U.S. flag painted on it, and the Control Center (with the firing rooms) is the low building to its right. The bays for the Shuttle Orbiter Processing Facility are in the center. In the distance is the runway added for Shuttle landing operations.

The Computer Miss Manners

Sometimes the pro-computer-checkout engineers went a little too far in needling their colleagues. In his book *Stages to Saturn* (published by the Government Printing Office in 1980), Roger Bilstein recounts the first automatic checkout of a Saturn IVB upper stage before a static firing test. The countdown progressed with extensive checking of equipment and components. Hundreds of engineers worked for dozens of hours, waiting for the order to fire.

A lead engineer typed the "go" command into his console. Nothing happened; the printer chattered out one line: "Say please." Befuddled by the unexpected response, he tried again. Again the computer said, "Say please." Finally he entered the command, followed by "please." The machine printed, "This is your friendly programmer wishing you good luck." Then the engine ignited.

Saturn S-IVB upper stage being removed from the static test stand.

The Computers-for-Cables Swap

In the '60s, NASA dreamed of launching many missions a year. This meant multiple vehicles had to be stacked, and multiple launch pads would be needed to handle the overlap in countdowns. A Vehicle Assembly Building (with a larger interior space than any structure in the world) was constructed next to a fancy blockhouse, the Launch Control Center. The Assembly Building had four bays, each large enough to stack a Saturn V, which stood over 360 feet tall. There were four firing rooms that could be connected to separate rockets for checkout.

The two buildings were very close together. The catch was that the two launch pads were (and still are) over two miles away from the firing rooms. Analog signals could not be transmitted that far without losses, unless complicated refreshing and amplifying took place. That process might have corrupted the data and increased costs. The only solution was to use digital computers. Digital data links passed discrete bits; parity- and error-correction software made these easier to check than analog signals. This arrangement dictated a computer in the firing room and another at the launch pad. Once a computer was available, using it to automate checkout procedures was a logical step.

Fortunately for the dissenting engineers, there was enough time in the flight programs to wean them from manual procedures. Computers and separate checkout cables stood side-by-side throughout the Saturn I flight program. By the 10th flight, 50 percent of procedures were automated. By the end of the Saturn V flight program, 90 percent were under direct computer control. Still, it took a 130-hour countdown to prep an Apollo-Saturn V for launching.

Stage Checkout

The Saturn V moon rocket was a three-stage, liquid-propellant vehicle—and the product of many different industrial contributions. The first stage had five engines of 1.5 million pounds' thrust each—still the most powerful single engines ever built. The second stage also had multiple engines, and the third stage, one. The stages were built by different contractors, such as Boeing and Chrysler; Rocketdyne built the engines. These contractors all had different ways of doing engineering; this meant each one had a different "language" for checkout. This was no problem as long as all the stages could be kept separate, but eventually they would have to be stacked.

Apollo-Saturn V moving to the pad on its mobile launcher. The 363-foot-high, three-stage vehicle has the Apollo Command Module just below the escape rocket at the top, and the Lunar Module under the tapered fairing.

Checkout procedures would have to be run again, often by different engineers.

Ludie Richard, a NASA engineer, led the development of a special language called ATOLL (Acceptance Test or Launch Language) to help make things easier. It had two advantages: it encouraged commonality among the checkout practices, and it could be used by the engineers themselves, avoiding the need to translate their requirements into terms a computer programmer (who might be inexperienced in rocketry) could understand. Over time, ATOLL coding caught on. In 1966, only six ATOLL programs ran in preflight checkout computers. By Apollo XI, the number had grown to 43 programs, and to 105 by Apollo XIV. *Test language* was gaining acceptance.

Testing the Spacecraft

Manned spacecraft checkout (especially of the Apollo Command and Service Modules) benefitted from the initiative of another NASA engineer. Tom Walton transferred to the Cape from a computer job at Langley Space Center. Assigned to Mercury checkout, he saw how difficult it was to monitor all the data points using cables. Every time the capsule had to be moved, the cables had to be disconnected and reconnected. A digital data link on a multiplexer would make things a lot easier.

Multiplexing

Sometimes it is not possible—or even necessary—to give every data link its own direct line. If data is sent sporadically, several sources can share the same cable. The data connections come together at a device called a *multiplexer*, which is essentially a switchbox like those used by the telephone company. The data links share the single connecting line, either on demand or by some time-sharing scheme.

He started small, first asking to build a digital ground station for telemetry. Each rocket and spacecraft had a telemetry transmitter that sent reams of data to Earth during flight. The data could be analyzed for faults later, but it also served engineers monitoring the flight in real time. They looked at analog dials, and strip charts like those from EKG machines. Walton got the go-ahead for his digital ground station, and started looking for a suitable computer.

After demonstrating that his scheme would work, Walton got the ultimate compliment: imitation.

Years later, a Shuttle begins the 3.5-mile trek to the launch pad on one of the same transporters used in the Apollo program.

Apollo Program engineers built a system called ACE (Acceptance Checkout Equipment) using two of the CDC-168s. One processed commands from control rooms and firing rooms. The other drove the displays in the spacecraft. When an Apollo sat on its launch pad, the ACE could still do its work over a nine-mile-long connection. But once the vehicle was stacked, the twin computers in the firing room and mobile launcher did most of the work.

Vehicle Checkout

The Saturn V made its majestic ride from the Vehicle Assembly Building to the pad on a mobile transporter with enough steel plate to build a light cruiser. Once in place, it was connected again to the automated checkout system. One hundred fifty thousand signals per minute raced along the data links. In a large room in the mobile launcher itself was an RCA 110A computer. Each of the four launchers had one, with a twin in each of the four firing rooms.

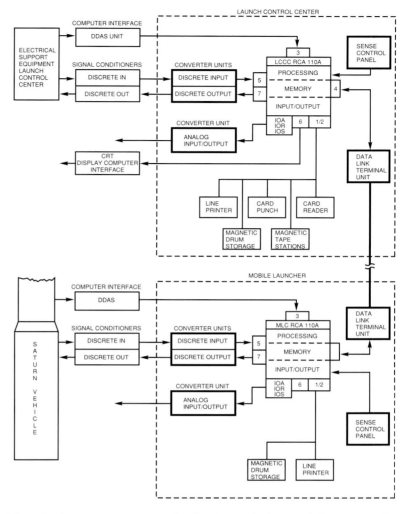

The dual computer setup in the Launch Control Center and mobile launcher.

Batch Processing
In the beginning of the computer era, a single program ran in a machine at any one time. When it was finished, a new program was loaded. In a way, the first computers were also the first "PCs"—one user per machine at any one time. As machines became faster and larger, programs were collected into a set called a *batch*, and loaded into memory. The programs in the batch ran one after the other, and then a new batch was fed in for processing. These batches were put together in decks of punched cards. During checkout of the Saturn, multiple tasks needed to run at nearly the same time, especially since there was only one computer for thousands of tests. Actually, it is amazing how much got done with this one minicomputer. It had only 32K of 24-bit-word memory. A big magnetic drum (a common alternative to disk drives in those days) held an additional 32K. Only half the memory was needed to hold the whole set of checkout programs, so they were duplicated to ensure reliability. Even then, they could monitor hundreds of signal lines.

The RCA 110A got its job because it was one of the first computers to have an *interruptible* central processor. This meant that a program running in the computer could be forced to give up its use of the central processor if something more important came along. If something needed immediate servicing, it was possible to write software that would save the memory location of the next step in the current program, plus all its data. After the reason for the interrupt was handled, the program could be restarted where it left off, with all its information intact. Nearly all modern processors have this capability; it is the basis for time-sharing and for operating systems.

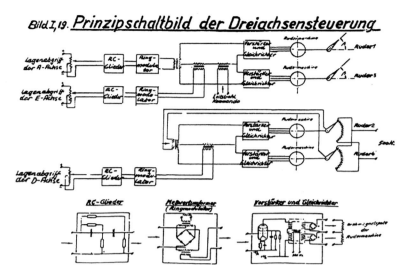

Circuits in the guidance computer of the A-4 (V-2).

On-board Help

One reason the RCA 110As were able to handle the work load is that they had a little brother aboard the Saturn to help. The guidance computer on the Saturn was the culmination of a quarter-century of effort aimed at improving missile guidance.

In Chapter 1, we saw how the Germans pioneered on-board active guidance systems on the V-2. The rocket actually followed a radio beam along the desired line of flight. If the rocket drifted to the left or right, changes in the radio signal strength caused the on-board analog computer to issue steering signals. That kept it on course. An engine cutoff timer and tilt table defined its range along this line of flight. Only a few modifications, none major, went into early-generation rockets built in the U.S. after the war. Even the Saturn I had a mostly-mechanical guidance system on its early flights.

It was the invention of the *inertial guidance system* that changed everything. An inertial system has a

stable member, kept stable by gyroscopes. Movement of the rocket around this stable member generates information that can be used in navigation by dead reckoning. This makes it possible to calculate the position and attitude of the rocket without the need for outside information (such as tracking by radar targeting).

Telemetry

Most experimental (or even operational) rocket flights have engineering sensors and other devices to measure the performance of the rocket and its payload (or human crew). Rocket scientists and technicians can study this data because it is sent by radio to ground stations. The stream of measurements is called *telemetry*. It is useful in reconstructing a mission in case of failure, and for determining what went right as well. Since the Mercury spacecraft had only about 100 measurement points, a low-cost PC could have done the job if it had been available, but at that time there were only two ways to go: mainframe or minicomputer. Mainframes of the time were still *batch-processing* machines; adapted to handle statistics and payrolls, they were ill-suited to computing in real time.

A V-2 rises from its test stand on the Baltic Island Peenemünde in the early 1940s.

A Different Era

Readers familiar with the modern PC industry may find it strange to hear about a computer made by RCA. In the 1960s, however, both RCA and General Electric (the electrical giants) had computer divisions. It seemed a natural extension of their electronic businesses. Both got big NASA contracts at the Kennedy Space Center. GE supplied two of its Model 635 systems for dual use: as batch processors for administration, and as display drivers for the blockhouses that would launch Atlas-Centaur, Delta, Saturn I, and Saturn IB. It was a weird combination of tasks, but it worked so well that the last 635 was not taken out of service until 1983, after 18 years on the job! By then, GE was long gone from the computer wars; Honeywell picked up its maintenance contracts and kept the old machines humming. RCA also left the business before the end of Apollo, but over a dozen RCA 110As stayed in use through the Apollo-Soyuz Test Project launch in 1975. (In 1983, with the help of a concerned NASA engineer, my research assistant and I salvaged some key components of the last 110A when its mobile launcher was in conversion to carry the Shuttle.)

The smaller engineering minicomputers were better for use in a changing situation. Walton settled on a Control Data CDC-168, a mini.

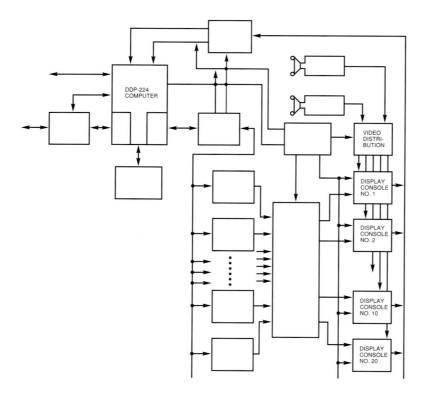

The computer-based display system in the Launch Control Center.

A sophisticated inertial guidance system was slowly incorporated into the Saturn program; a computer was added to take advantage of it. On the fifth flight of the Saturn I, an ASC-15 computer (built by IBM for the Titan II intercontinental ballistic missile) went along; the new system ran in parallel with the inertial guidance system already installed in the Saturn. For flights six through ten, the on-board ASC-15 was the primary guidance computer.

While the Saturn I rockets finished their flight program, the S-IVB upper stage was readied for use as a second stage on the Saturn IB, and as a third stage on the Saturn V. The Saturn IVB had a relatively thin ring of skin attached, within which sat the guidance system and a new computer: the Launch Vehicle Digital Computer, or LVDC.

The LVDC's design built on the past and looked ahead. It used some technology from the

Gemini spacecraft guidance computer (see Chapter 5), and pioneered some that showed up later in the Skylab Apollo Telescope Mount Digital Computer (see Chapter 7). From Gemini, it borrowed memory modules of 4096 words (each word 26 bits long), each module in duplicate for reliability. As they would in Skylab, triple-redundant modular circuits "voted" on all signals passing through them (for more on the use of redundancy in space flight, see Chapter 7).

Obviously, the designers were keenly interested in making sure this computer would not quit. It actually had a relatively short life (from six seconds before liftoff to about 6.5 hours into a lunar mission), but every moment was a busy one. Most important for the checkout engineers, the computer could be used as a sophisticated multiplexer that connected to other spacecraft systems. Therefore, the LVDC was not only a subject of checkout procedures, but a helper.

The Ultimate Labor-Saving Device

In the firing rooms, in addition to the RCA machine like the one in the transporter, there was a DDP-224 computer for displays. (These were just like the ones we met in the last chapter driving simulators.) Over 400 engineers crammed into a firing room for a Saturn V launch. There were over 100 display monitors, and a host of analog meters and strip printers. As elaborate and complex as launch processing had become, it still represented a tremendous saving in time and effort.

Think about this for a moment: if 400 engineers (there were more in other rooms doing other jobs, but for now, let's stick with just this number) worked thorough a 130-hour countdown, they would run up 52,000 hours of labor. A typical work year of the time was 2,000 hours, so it took *26 person-years* to do final prep on a Saturn V. This was *with* automation, and does not figure in the time spent on stage and vehicle checkout at manufacturers' sites and in the Vehicle Assembly Building. If all this work had been done by hand, building and readying a moon rocket would have been more like constructing the Great Wall of China. Generations of families of engineers would have had to live and work on the Space Coast in Florida to get one rocket off the ground. The Shuttle would have required even more work.

Automated Launch Processing for the Space Shuttle

3

It is July 16, 1969. You are standing in the back of one of the still-new firing rooms, having come out of the computer area where you nervously checked (for the hundredth time) the switch positions on the main control panel of the RCA 110A. In front of you is a sea of gray consoles and white-shirted engineers. Eight rows with a couple of aisles, then five longer rows with one aisle, then three tapering rows. Visitors sit up on the right, the operations managers on the left. Von Braun, the leader of the German rocket team, is there—a long way from Peenemünde.

Through the windows with their blast shields (if a Saturn really managed to fly straight into them, would the blast shield matter?) you can see the big rocket with Armstrong, Aldrin, and Collins in the seemingly-tiny tip. Work shifts or no work shifts, some senior people just would not leave for any length of time during the days of the countdown. You think you can tell them from the luggage under their eyes, but nobody is looking too good right now.

Space Shuttle Endeavor in the Vehicle Assembly Building, Kennedy Space Center.

The crowded firing room in the Saturn Launch Control Center.

Finally, the count reaches zero. Seven-point-five million pounds of thrust push the Saturn V up from the pad; ice on the outside of the cryogenic fuel tanks cascades off, in the biggest avalanche Florida has probably ever seen. The rocket is rising nearly three miles away, but you can *feel* it shake the building. The crowd outside is going nuts. Inside, you can sense the combination of fear, relief, pride, and a little emptiness. Apollo XI is racing to the moon, but now Mission Control in Houston has the helm. Hundreds of engineers wait, temporarily unneeded, looking forward to watching the progress of the mission just like everyone else: on television.

Shuttle as an Airliner for Space

That was the historical reality; here is a futuristic dream from 1969: 10 years after Apollo XI, you watch a typical Space Shuttle mission end with a landing on a nicely-paved long runway at the Cape, only a few yards from the firing room where you saw it start. The winged spacecraft is towed directly into a horizontal hangar, the Orbiter Processing Facility. A big cable connector is plugged in where the umbilical cord at the launch pad once went. At the other end of the cable are a small platoon of minicomputers in a firing room.

Automatic preflight checkout begins for the just-returned Shuttle. The firing-room computers talk things over with the on-board computers (each of which has tentacles into the spacecraft's most complex components, those of its avionics system). Meanwhile, a dozen or so technicians inspect the heat-resistant tiles, reaction control system, and brakes for wear and tear. A new payload is installed in the cargo bay.

Inside the Vehicle Assembly Building, now nearly empty (a Shuttle is less than two-thirds the height of a Saturn V), a reusable winged booster waits on its mobile transporter. It got the mission off to a jump start several days ago, and returned to base minutes after separating from the orbiter. By now

it is refurbished, inspected, and ready to go, its own computers quietly exchanging information with the firing-room computers.

At last the orbiter emerges from the Processing Facility, and is towed across the road to the Assembly Building. A big gantry crane lifts it up

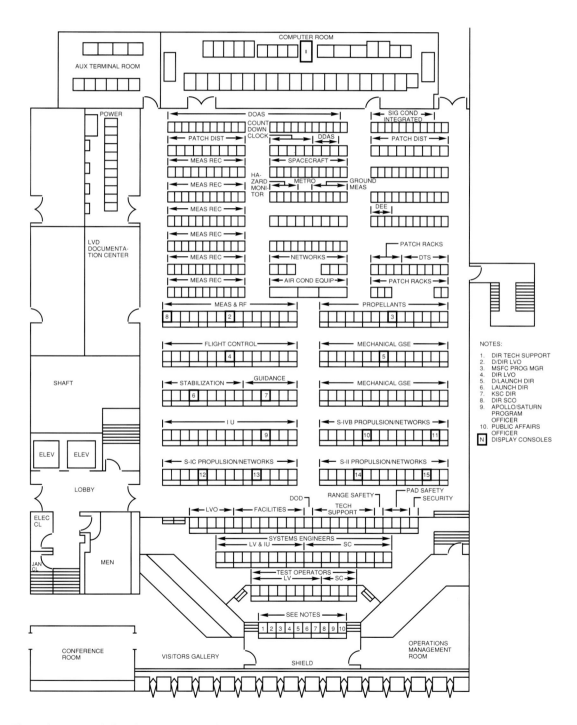

Floor layout of the Saturn Launch Control Center.

to a vertical hang. It is mated with the booster. Now the orbiter's computers communicate with those of the booster, and with the checkout computers in the firing room.

Maybe 12 days after landing, you see the checked-out space vehicle rolled out to the pad. Two days later, the astronaut crew enters the flight deck at T-minus two hours. They lie in their acceleration couches, silently monitoring the computer screens as the 120-minute countdown progresses. In the firing room, a couple of dozen engineers are also watching screens. You find it hard to believe the size of the empty spaces between the computer consoles. Only the VIP seats up front are still crowded. You miss the handsome von Braun. When the computers agree all is well, the Space Shuttle hurries off on the 46th of its 75 missions planned for this year.

Shuttle as an Infrequent Flyer

That was the dream. Here is another historical reality, this time from 1989: it is twenty years after Apollo XI. The 31st Shuttle mission overall, STS-34, rises into space to deliver *Galileo* to its rendezvous with Jupiter. In the first six full years of flights (skipping the two-year hiatus after the *Challenger* disaster) just about five missions a year have flown.

With the current launch-commit criteria, flight rules, paperwork, budget, and "flow" (the things that happen to a Shuttle preparing for flight), it is a major achievement if NASA ever hits 12 Shuttle missions in a year. There is still an enormous amount of sheer physical labor and paperwork in preparing a Shuttle for flight.

In retrospect, bringing airline maintenance folks to NASA to give advice during Shuttle planning was probably a waste of time. But one part of the

The Shuttle lands horizontally on a runway. Note the drag chute that helps decelerate the Orbiter.

dream did come true: Except when an orbiter is towed from the Processing Facility to the Assembly Building, or a stacked vehicle is in motion out to the pad, there *is* a direct computer connection to the assigned firing room. Engineers can monitor what is going on in the spacecraft at all times, even when someone is working on it. There is a wealth of information about spacecraft health available to everyone involved. This automated system is a far-advanced descendant of the one used in launching the Saturns.

Once the crawler leaves the Shuttle on the launch pad, the spacecraft is again in computer contact with the firing room.

Thinking Ahead to Shuttle Checkout

The original Shuttle concept—using both a winged booster and winged orbiter—meant it could have been launched from any number of sites, mostly inland. Under budgetary pressure, NASA finally decided (in early 1972) to use two solid-rocket boosters instead, and recover them by parachute. All but the coastal launch sites had to be ruled out; it was better that the boosters come down in the water. There, relatively undamaged, they could be towed back to shore for reuse.

Since NASA wanted to have assembly, checkout, and recovery operations in the same place, Kennedy Space Center—an existing coastal launch site—was a natural choice. Relatively speaking, it did not take much money ($335 million or so) to convert the old Saturn Assembly Building and mobile launchers, build the Processing Facility, pave a runway, and convert the pads. The item that really set the pace of the conversion was a completely new checkout system. It would integrate the old "stage" and "vehicle" procedures.

Distributed Computing

Sometimes so much data of different types must be processed that it is better to have several individual computers connected to different parts of a system. If a single mainframe computer had to handle all the inputs of the Shuttle checkout system, it would be hopelessly slowed, and its software unnecessarily complex. Dedicating a computer to each major subsystem (such as propulsion or avionics) lets the computers run their processes *in parallel*, instead of one after the other. This makes the checkout more efficient.

NASA had an internal competition to see which of its centers would have the job of designing and procuring the new computer system. Johnson

Distributed Computing in Preflight Checkout

Inside the Vehicle Assembly Building at Kennedy Space Center during the Apollo program. The mock-up Command and Service Modules in the foreground were used to practice stacking a moon rocket.

One of the more experienced NASA engineers, Frank Bryne, set the architecture of the new Kennedy Space Center system. He didn't like the "monolithic mainframe" approach. Using one computer, the checkout programs would most probably have to be run serially, which would stretch out the time required for countdown. Bryne thought a collection of minicomputers would be much better. You could run things in parallel wherever possible, speeding up the countdown. Also, this *distributed computing* approach let each major component engineer have his own processor and memory to play with. It would not be necessary to fight some "central memory allocation committee" for working space in the central computer.

The Kennedy team's original concept for the new system remains in use today. They decided early on to have three major subsystems in the overall Launch Processing System; that decision set the computer requirements. A cluster of minicomputers in each firing room—the Checkout, Control, and Monitor Subsystem—would connect directly to the space vehicle and run the checkout programs.

The minicomputers were programmed in a new checkout language called GOAL (Ground

Space Center had a team working the problem. They recommended a single central computer with a battery of display consoles, much like the RCA 110A setup in the Saturn firing room. Meanwhile, the Kennedy Space Center assembled a team that aggressively sought the job. To demonstrate what they could do, they put together a model of a hydrogen tanking facility. Using a Digital Equipment Corporation PDP-11/45, they made a neat graphic display of the entire system. The computer was connected to the model, which actually had moving parts. Schematics showed pipes, tanks, and valves; critical data was displayed. The NASA higher-ups were very impressed. Kennedy Space Center got the job.

Operations Aerospace Language), the successor to ATOLL. Much more readable than ATOLL, GOAL used statements like READ, IF-THEN-ELSE, and VERIFY. The engineers doing the preflight work could write their own GOAL programs, and test them against simulations of the vehicle kept in a special mainframe computer, the Central Data Subsystem (CDS). Once satisfied that the GOAL code was correct, the engineer could store it in the CDS until needed.

In addition to these two subsystems, there was a third component: the Record and Playback Subsystem. Like a giant flight recorder, it would save all telemetry data from the spacecraft, throughout the mission. An engineer could call up information from a few days (or tests) ago, and compare it to current readings. Entire flights could be reconstructed.

Such a distributed network of computers and software meant solving some hard interconnection problems. This was before standard local area networks like Ethernet were in common use. NASA had to be innovative.

A Post Office for the Computers

There is a saying in the data-processing industry: "Two copies of data equals *no* copies of data." Trying to maintain two perfectly equivalent (but physically separate) databases is a nightmare. If any component in either chain fails at any time, the contents of the database begin to diverge. At that instant, neither one is an exact copy of the other.

Nearly lost in the scaffolding in the Orbiter Processing Facility is the Shuttle *Columbia*.

Moving to a distributed system to preflight the Shuttle forced NASA to create a common database for its various subsystems. In previous checkout systems like ACE, a portion of memory of one machine acted as a common data location for the others. In the case of the Launch Processing System, one machine's memory would not be enough.

Local Area Networks

In most well-equipped offices today, the PCs are interconnected in what is called a *local area network*. This means the computers that work for the people in a department or division of a company can share data (and sometimes software) across a commercially-available connector. Often these networks (*LANs*) are connected to other LANs in different sites, forming a *WAN (Wide Area Network)*.

Byrne decided to build a customized *data buffer* as the heart of the system. This idea went against the

decision to use off-the-shelf commercial equipment for the hardware, but no commercially-available device could do the job. Even so, the internal workings were bought from standard parts suppliers; the *design* was different—so different, it was even patented.

The Common Data Buffer, as it came to be called, was designed as a secure, incorruptible place to put information. As such, it had to attain extremely high reliability in both its hardware and software.

The Common Data Buffer had an unusual architecture. Most of the time, computer words would be stored in a memory bank whose measurements were the width of a word and the length of the memory (for example, 64K of 16-bit words). The Buffer memory, in turn, was subdivided into 32 *blocks*; each block was one bit wide, and measured as long as the memory (64K). That way, any time a block failed, only one bit of any word would be lost.

Since its error-correcting codes could correct 100 percent of such one-bit errors, the Buffer was an extremely reliable storage device. Even so, the entire memory was duplicated—making it even more reliable.

The 32 one-bit-wide blocks of the Buffer's memory were designed to accommodate computer words that were only 16 bits long. As each word was transmitted in the system, it would be split into halves, with 16 bits of error-correction code inserted between the halves. The first half of the word was followed by eight bits of correction code. The second half of the word was preceded by eight bits of correction code. This "overkill" was intended to compensate for the big signal losses a message was expected to suffer as it crossed the network. Fortunately, such signal losses have not occurred much at all over the years.

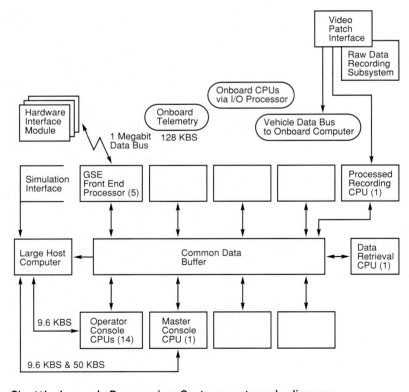

Shuttle Launch Processing System network diagram.

Launch Control Center firing room, reconfigured for the Shuttle.

Each firing room has its own Data Buffer, connected as shown in the figure. At lower left in the diagram, two interconnected boxes represent the big computers in the Central Data System, and the 14 minicomputers used by the checkout engineers. Notice that the data transfer rate is 9600 baud—just about what you can get on a decent telephone line today when you use a modem to log into a remote computer from your home machine.

Every device shown in the diagram was connected to the Common Data Buffer in some way. Each data item needed for preflighting the Shuttle had its own "mailbox" in the Buffer's memory. If an engineer running a test from his console needed access to the data, he could get it without necessarily cutting anyone else out, and his computer would "see" the same data as any other device on

the network. While a particular computer was updating the data, all others were locked out from changing it.

Old Reliables: Minis and Mainframes in the Launch Processing System

The diagram of the Shuttle Launch Processing System shows how its various components were connected. You can see the devices for the Record and Playback Subsystem in the upper right, the Central Data Subsystem to the lower left, and the Checkout, Control, and Monitor Subsystem across the bottom.

There was also a row of "front-end processors" (pictured above the Data Buffer); these connected

```
180:1857/40 INTGB                                                          ⬛
SM SA 0123 D123 E123 1AB 2AB 3AB 4AB 5AB 6AB $SYS                    F123456
TCGS0
55/08 OSTMCN1225 COMMAND VALID ONLY FROM CONSOLE WITH SYS. INTEGRITY
                            ──── SYSTEM STATUS ────
 S N  BS LP C L C G S    S N  BS LP C L C G S    S N  BS LP C L C G S
 T A  UL 00 F O O O W    T A  UL 00 F O O O W    T A  UL 00 F O O O W
 A M  CO GR I A O O I    A M  CO GR I A O O I    A M  CO GR I A O O I
 T E  KT  T G D M M T    T E  KT  T G D M M T    T E  KT  T G D M M T

 □ GS1A 00  1 * - -   - I  □ C1   07 22 * - -   -   □ TCG1 0F 42 * * * * A A
 □ GS2A 00  2 - - -   - I  □ C2   07 23 - - -   -   □ TCG2 1F 43 * * * * * S
 □ GS3  00  3 - - -   -    □ C3   07 24 - - -   -   □ TMDS 2F 44 * * *
 □ GS4  00  4 - - -   -    □ C4   07 25 - - -   -   □ VDA  3F 45 * - -   -
 □ GS5  00  5 - - -   -    □ C5   07 26 - - -   -   □ GS1S 15 46 - - -   -
 □ GPCA 22  6 * * * * I    □ C6   07 27 - - -   -   □ GS2S 00 47 - - -   -
 □ OIA  22  7 * - -   - I  □ C7   07 28 - - -   -   □ GPC5 02 48 - - -   -
 □ SWOF 3F  8 - - -   -    □ C8   07 29 - - -   -   □ OIS  22 49 - - -   -
 □ SWDF 3F  9 - - -   -    □ C9   07 30 - - -   -   □ LDBS 00 50 - - -   -
 □ ET   02 10 - - -   -    □ C10  07 31 - - -   -
 □ ME1  02 11 - - -   -    □ C11  07 32 - - -   -
 □ ME2  02 12 - - -   -    □ C12  07 33 - - -   -
 □ ME3  02 13 - - -   -    □ MSTR 2A 34 * * * *
 □ SDB1 15 14 - - -   -    □ INTG 07 35 * * * *
 □ PLD  16 15 - - -   -    □ BKUP 07 36 * - -   -
 □ SDB2 00 16 - - -   -    □ ESA1 31 37 * * * *   □ ESA2 14 57 - - -   -
 □ LDBA 15 17 * * * * I    □ PDR  0A 38 * * * * I
 □ HOSC 02 18 - - -   -    □ SPA  1B 39 * * * *   □ ESD  2E 59 * * *
 □ UPLK 05 19 * * * *      □ PP1  00 40 * * * *
                          □ PP2  15 41 * - -   -
 □ LDBD 35 21 - - -   -
 □ ECC                    □ REF  DES                              □ TERMINATE
```

Typical display in the Launch Processing System.

to the spacecraft in various ways, and processed data from various sources. For example, the Ground Support Equipment processors had a direct hardware hookup to the Shuttle through its HIMs (hardware interface modules). Pulse Code Modulation processors received data from the Shuttle's telemetry transmitters. The HIMs could exchange data at a relatively high rate (a megabit a second), while the telemetry rate was about an eighth of that.

Data Buffers

Whenever two computer systems of different speeds are interconnected, it is necessary to have a *buffer* between them. Data is stored in the buffer temporarily to compensate—to help the faster system slow down, or the slower system build up a block of data for fast transfer. The Common Data Buffer performed in that role, and also provided the firing-room computers with *mailboxes* (storage locations for share data).

The firing rooms were dominated by blue consoles (no more institutional gray). Some contained 16-bit minis built by the Modular Computer company. Three display terminals were connected to each ModComp computer; the total made up one console. For program storage, the computers had disks with capacities that ranged from five to 80 megabytes, and 64K of memory. Most GOAL code was still kept in the CDS; it was the only code needed to run a particular console.

The word size and small hard drives of the computers may seem primitive today, but the displays were really advanced, even for today. There were liberal numbers of function keys, and mouse-like cursor control. The displays were in color, unlike almost every other display monitor in the Shuttle program. The terminals could also be switched away from their consoles' ModComps, to function as terminals for the hardware in the CDS. The front-end processors were also ModComps, some with bigger memories. If you stood in the middle of the room and counted, you might see as many as 30 minis humming away. This was not a small network for the mid-1970s.

Downstairs from the firing rooms in the Launch Control Center were the dual Honeywell 66/80 mainframe computers that made up the CDS. Originally they had 500K of memory for their own processing and 500K of shared memory the four of them could access. As the GOAL programs proliferated, NASA upgraded them to 1.5 megawords, with one meg of shared memory each. The initial installation of 172 disk drives had a capacity of 30 billion bytes.

Walking into the rooms where these disk drives were housed was—and is—an amazing experience. Even with all the technology surrounding the space program, that feeling remains as close as you can get to being in a science fiction film. It is difficult to believe any one operation could own that many disk drives.

Over the years, NASA and its integration contractor for the Launch Processing System have changed Honeywell's operating system so many times that Honeywell could no longer maintain it. Even so, *every single piece of the Shuttle checkout system is operative today, exactly as it was when it ran for the first time in 1979.* In a typical configuration, a pair of mainframes supports a firing room, while another pair is used for software development. Each machine runs self-tests that help it decide whether it is still working right. If it does not "think" so, it passes control to its partner.

History and reality may not have reflected the 1969 dream of everyday space flight, but NASA continues to make thorough use of its equipment. Even though the Shuttles are not launching at anywhere near the rate envisioned in the 1960s, the processing facilities at Kennedy Space Center are far from underutilized. Orbiters are prepped for their next missions as quickly as possible.

The pervasive automation means far fewer technicians are needed in checkout work, but many labor-intensive procedures remain. Currently, for example, the heat-protective tiles are tediously examined and manually injected with a toxic waterproofing chemical—thousands and thousands after every flight. In response, NASA has tried to innovate even more. The Applications and Simulations Branch—supplier of the vehicle simulations against which engineers test their GOAL programs—explored using artificial intelligence to make suggestions when anomalies came up. A prototype robot, the Tesselator, is being built to inspect and waterproof the Shuttle's tiles.

But NASA is organized for the business of manned space flight in a way that goes beyond the Kennedy Space Center. The ground-based control of the missions is transferred from the launch site to the Johnson Space Flight Center at the moment of liftoff. On the unmanned side, deep-space probes (those flying beyond the point dominated by Earth's gravitation) are controlled through project-specific centers, most often at the Jet Propulsion Laboratory in California. JPL served—and still serves—as headquarters for the Earthbound masters of *Galileo*; formerly the "passenger" of STS-34, the probe continues its flight to Jupiter. Both Johnson and JPL have an active tradition of innovation in their use of computing power to control complex space missions (and for unmanned spacecraft, these are often lengthy). Their story is next.

Mission Control Centers 4

"Go for throttle up!"

"Roger, Houston, go for throttle up!"

Atlantis commander Don Williams continues the litany spoken by so many of his predecessors. In a surprisingly small room in Building 30 at the Johnson Space Center, technicians sit at consoles studying data readouts from the Shuttle's telemetry. One is saying the names of the critical events in the ascent to the crew, at the times they are supposed to occur. If all the telemetry data is in the green, then the call "go" is permission to continue.

Williams should see the Shuttle main engine thrust readout move to 105 percent of rated power after "go for throttle up." After the initial climbout and roll, engine thrust is reduced to ease the passage through the "max-Q" region where stresses on the spacecraft are greatest. Clearing that, the engines go to beyond "full" thrust until cutoff.

Manned Spacecraft Center, Houston: Mission Operations Control Room.

The checking and rechecking of components does not end with liftoff. Mission Control engineers continually monitor the health of the spacecraft and its crew. They also keep track of the meticulously-planned sequence of events. Each mission is scripted, rehearsed, and then executed. Astronauts follow the script and its contingency plans during training, supported by the controllers. If anything goes wrong, like the frightening explosion on Apollo XIII, the team redoubles its support.

From the very beginning, digital computers have been the heart of Mission Control. Controlling missions remains the only aspect of the manned space program that can claim such a long-standing partnership with the computer. Also, IBM is the company most closely associated with the Mission Control Center's Real Time Computer Complex. IBM mainframes have always been the computer of choice. In the process of supporting the Real Time Computer Complex, IBM developed a number of devices and software innovations that carried over to its commercial business. Its success with these is one of the few technological side effects of the space program that is both beneficial and easily identified. In December 1993, IBM sold its Federal Sector Division to Loral. This was the division which supported the space program, both on the ground and on board flight vehicles.

It is difficult to believe that the IBM officers responsible for the sale fully realized their division's performance and potential—especially the technology it had generated and transferred.

The Early Days: Vanguard and Mercury

One of the big science initiatives of the 1950s was the International Geophysical Year (actually about 18 months in 1957 and 1958). The United States offered to put up an artificial Earth satellite as one of its contributions to the effort. The Army and Navy submitted competing proposals for projects aimed at launching the satellite.

The first launch of the Space Shuttle, April 12, 1981.

For political reasons, the Army lost out, even though it was much further along technologically. The main stage of the Army launcher was the Redstone, a medium-range ballistic missile much like the infamous Scud used against Israel and Saudi Arabia by the Iraqis in Desert Storm. In order to test re-entry nose cones for intermediate-range and intercontinental ballistic missiles, the Redstone boosted a package of solid-propellant stages into the upper atmosphere, then the stages turned toward the Earth and fired, accelerating the nose cones to 17,000 miles per hour—the speed an ICBM would reach on re-entry. The Army team, led by Wernher von Braun, figured out that instead of aiming the solids at the ground, if they were fired toward space, the addition of one more stage would make it possible to attain orbital speed.

The political problem was that the Redstone was a weapon. The Eisenhower Administration thought it better to have a purely scientific launcher. The Navy's proposal also called for a multistage vehicle—but its first stage would be a Viking sounding rocket. Here was one of the first large, home-grown, liquid-propellant rockets used for atmospheric research; it was "politically correct" for 1957. Project Vanguard got the nod; so would IBM.

The first satellite Vanguard would launch was tiny (3.5 pounds), and might be hard to find in the sky; collecting and analyzing its data was a concern. The Project set up a Real Time Computing Center down the street from the White House in June 1957, about six months before the first launch attempts. An IBM 704—one of the more

Because of repeated Vanguard failures and pressure from Sputnik, America's first successful satellite was launched with a Jupiter-C rocket. Some later Vanguard flights were successful.

successful early mainframe computers—was the heart of this system. It calculated the orbital path of the Vanguard so ground stations would know where to point their antennas to get the data. It also helped with data analysis.

Ironically, the Army rocket beat Vanguard into space. When the Soviets shocked everyone with the Sputnik in October 1957, the Army was told to go ahead full speed as a backup to the Navy. The first Vanguard blew up on the launching pad in December 1957. In January 1958, the Army vehicle—the Jupiter-C—orbited a satellite.

In 1959, the newly-formed NASA started to think about a ground-control center for its equally new Mercury manned space flight program. Western Electric got the contract for the tracking system and ground stations. IBM, fresh from its Vanguard experience, received the hardware and software subcontract for Mission Control. They were able to use their existing computer center in downtown D.C. for about a year.

Redstone missile test.

A Rocket by Any Other Name . . .

The first stage of the Jupiter-C satellite launch vehicle was a Redstone ballistic missile like the one illustrated here. The stubbier Jupiter was one of the first intermediate-range ballistic missiles. So why confuse the issue by using the name "Jupiter" with a Redstone-based vehicle?

Apparently using the Redstone with a cluster of Sergeant rockets as upper stages could not be funded with Redstone development money. There was, however, plenty of cash in the Jupiter budget. The *Jupiter-C* got its name from a need to associate the source of funding with the missile.

The General Accounting Office probably had a great time trying to figure this one out.

Jupiter missile launching.

In 1960, the Goddard Space Flight Center came into being outside the Beltway in Maryland, about a half-hour drive from central Washington. NASA bought two IBM 7090 mainframes as central computers for Mission Control; they would be located at Goddard. However, when IBM first set them up, the building in which they would be housed was not yet finished. So a makeshift structure of plywood, canvas, and window-mounted air conditioners surrounded the computers. This was a far cry from the typical computer installation of the time, with its raised floors, controlled air, and near-operating-room cleanliness.

The 7090 was a good choice for the space program. One reason it was developed was to respond to a Defense requirement: a machine to process data from the new Ballistic Missile Early Warning System. The sorts of data a missile defense system and a suborbital (or orbital) space flight would generate were a close match.

The Mercury computer network actually had five machines. Two were in Maryland at Goddard: the Mission Operational Computer and the Dynamic Standby Computer. If the primary failed, the standby took over (which actually happened a few times). Two were at the Cape, where the controllers sat. One of these, an IBM 709 for trajectory calculations, gave data to both the controllers and the range safety officer; it continuously updated the impact point of the booster. The other computer at the Cape was a Burroughs/GE system that would provide guidance for the Atlas booster. It sent commands by radio (unlike the guidance computers for the Saturn, which would ride along). The fifth computer was another 709 at the Bermuda tracking station. It also did trajectory calculations, and could take over the mission if the dual computers failed at Goddard.

With controllers and computer separated by a thousand miles, there was a great need for improved communications and maximum processing speed in the Mercury system.

Breaking the Bottleneck

IBM implemented two innovations for Mercury. One increased the *throughput* (processing speed). Most computers of the 1950s had a sort of speed limit shared even by today's computers: the "von Neumann Bottleneck."

John von Neumann was one of the scientists who proposed the coding of software in the same form as data, thus making possible the use of stored programs in computer memory. At the time, "programs" were actually external to the heart of the machine. Charles Babbage, an English computer architect of the 1800s, had his programmable machines designed to run under the control of punched cards. As the instructions were read off of the cards, they were executed.

A Mercury-Redstone spacecraft is readied for launch at Cape Canaveral, but its flight will be controlled from a thousand miles away.

In the 1900s, accounting machines, many built by IBM, were programmed the same way. Even ENIAC, the first fully electronic digital computer, had its program steps on cards. ENIAC, however, kept its initial data electronically in a special rack of *registers*; these were circuits that could be set by hand using rotary switches. Putting programs in electronic memory was a great step forward; computers became more flexible and usable. It also meant, however, that the connection between memory and central processor had became a critical factor in determining speed.

An IBM 7094 computer system like the ones used in the Gemini Real Time Control Center.

In addition, input and output passed through the *CPU* (the central processor, which also had to run the machine) on their way to and from the memory. Data could be slowed in a traffic jam of signals. The potential bottleneck was named after John von Neumann.

IBM got around the bottleneck by inventing the *data channel*. Big blocks of data could be entered directly into pre-assigned memory locations where the software could manipulate them. That way the CPU would not have to suspend processing in order to handle input.

Even though the speed problem was more or less solved at Goddard, the long communication line to the Cape slowed things down again. The data link ran at a blazing 1000 baud! (Nowadays even the worst U.S. telephone line permits faster data transfer, through even a cheap modem.) Some displays in the Mission Control Center lagged behind the real flight situation by up to two seconds—somewhat dangerous for a vehicle that could travel *10 miles* in that time.

From One Process at a Time to Many

IBM's second innovation was in the "operating system" of the computer—though that contemporary term is too sophisticated for the actual software-control tasks that were possible then. Most computers of the time were still "batch" processors. A program ran until it was done, then another program was loaded, and it ran until it was done.

To help move things along, some of the newer computers came with "monitor" programs. These sought to take advantage of the 32K memories (considered huge) that were becoming common. Several programs would be loaded into memory from card decks and executed in sequence. The monitor program kept track of this execution. If a program exceeded its allocated time (or failed in other ways), the monitor kicked it out and moved along to the next program.

This innovation soon proved useful; simple batch processing would not work in a complicated space

Gemini spacecraft launched atop a Titan II.

mission. Programs did not run completely in serial. Some needed a few milliseconds of processing, then they could wait awhile for others to run—in effect, they could take turns. IBM invented the "Mercury Monitor" to handle this situation. The programs needed for any particular mission phase were separated into batches; a particular event would initiate each one. At liftoff, for instance, a pre-set collection of programs would be running in the computer, each using its share of CPU time. They would be loaded automatically from magnetic tape as needed.

The Mercury Monitor program was actually doing much more than just monitoring; it also handled the interrupts caused when the CPU had to process high-priority data or program requests. This type of system predated (and heavily influenced) the *time-sharing* software IBM would offer

its mid-1960s commercial customers with the System 360 computer.

As a Mercury astronaut flew around the Earth, the controllers on the ground would monitor the spacecraft systems. Throughout the mission, the computers were in use—providing continuous updates on time to retrofire, as well as other guidance and navigation data. The controllers had to decide within seconds of engine cutoff whether an orbit had been achieved, or whether to put the emergency recovery forces in motion. After a few flights, this became a fairly smooth operation. The computers' improved speed had made a vital difference.

As the Mercury Program wound down and the two-man Gemini Program spun up, plans had to be made for controlling much more complex missions (see the next chapter). The Mercury Mission

An IBM System 360 computer.

Apollo mission controllers watch the final descent of the Command Module to its water landing.

Control system was adequate for the first few Gemini missions; these were simply two-man extensions of Mercury flights, with little orbital maneuvering. NASA had decided, however, to centralize the Mission Control functions done by Goddard and the Cape. Now both the computers and controllers would be moved to Houston's new Manned Space Flight Center. For the last 30 years, that is where the Mission Control Center—and its associated Real Time Computing Center—have remained and operated.

Mainframes and More Mainframes

Based on what was learned during Mercury, NASA increased its mission requirements for Gemini. Changes in the mission-control systems would be necessary. For Gemini, they would have to control at least two spacecraft at once. NASA also wanted to use more graphic pictorial displays, instead of endless text screens that made it hard to tell what was really important. Finally, they

wanted even better reliability: a 0.9995 probability of continuous computer support. This goal was ambitious; in those days, hardware failures dominated software failures as the chief cause of downtime. If the desired reliability were converted to a batting average, the player would fail to get a hit only five times in 10,000 turns at bat.

IBM won the contract (surprise!) for the Real Time Computing Center's hardware and software. They started with three new 7094 mainframes, each with 32K of main memory and 98K auxiliary memory—giants of the time. The 7094 was an upgraded version of the 7090 model, with a better operating system. Eventually the computer center would be further upgraded to *five* advanced-model 7094-IIs, each with 65K main memories and a half-megaword of fast auxiliary memory.

The main difference between the main memory and auxiliary memory was that the computer could address the main memory directly; the auxiliary had to do some machinations inside the CPU to access it. It was still a lot faster than magnetic tape!

The slightly crowded Mission Control Center for the Shuttle flights.

The five 7094-II computers had separate jobs. As in Mercury, two were set up as the Mission Operational Computer and Dynamic Standby Computer. Two others simulated both the control network and the ground operations, so that software in the mission computers could be tested. The fifth was the software-development machine.

Over 400 programmers worked on developing software for the Real Time Computing Center. They used various development tools—some custom, some off-the-shelf. The 7094 came with an operating system called IBSYS, which included some basic tools: compiler and editor programs. The language was FORTRAN (already suited to high-level programming), augmented by some special code. The programmers were quickly able to add advanced features to the mission-control software.

An Executive program of 13,000 words replaced the Mercury Monitor, and gave the system interrupt and real-time capability. To speed up processing, the programs for the next mission segment would be loaded from tape to the auxiliary memory. That way they could be moved quickly to the primary memory for execution. (This "look-ahead" capability and use of a fast buffer gave IBM some good ideas for its later virtual-memory computers, the 370 series.)

Apollo Mission Control Computers

Even with these hardware and software enhancements, the 7094s began to fill up and slow down. By this time IBM had announced its System 360 line. These new business computers had a multiprocessing capability—partly due to technology

Is It Real, or Is It Virtual Memory?

Processors are often faster than the memories that serve them. One solution to this problem is the use of *virtual memory*. Most frequently, virtual memory is implemented by putting only one "page" of a program (and its related data) in "real" memory, and putting the pages next-most-likely-to-be-needed in a special place on a fast disk drive. In this way, many users are sharing the fastest memory, which gives them the illusion that each of them has complete control of the computer. In actuality, most of their data and code is residing on a disk, much like an actor waiting for a cue to rush out onto the stage.

The *auxiliary memory* in the Real Time Control Center was additional fast memory added to what were considered, in their day, "giant" computers. Auxiliary memory acted as the "off-stage" waiting area for programs stored on magnetic tape (which was hopelessly slow for real-time applications).

transfer from the space program—that might restore some speed. NASA initially wanted to migrate (change over) to the new machines, but IBM had announced the new line of computers long before it was ready to ship them. In the meantime, Control Data had released its 6600 line—computers faster and (by 1965 standards) somewhat more capable than the 360s. Highly-placed NASA officials wanted the Mission Control Center to buy the 6600s and let IBM keep the software contract. IBM countered the threat to its business by shipping the first available 360 to Houston. The new machine replaced the software-development 7094 as NASA began preparations for the long, complicated lunar missions of Apollo.

It turned out that extensive software modifications were needed to handle the new missions. The interrupt capability on the 360 was not robust enough to handle the increased demand. IBM modified its new operating system (OS/360) into RTOS/360, the Real Time Operating System/360. The adaptations took time. The FORTRAN code from the Gemini days moved over to the new computers more or less intact, but various other fixes—including upgrades in the machine code—lengthened the process. When they finally arrived, the 360s had about four times the memory of the 7094s—but they filled quickly with new functions, and increasingly the programmers used tricks and "hacks" to make things fit.

Despite the use of such "artistically made" software, the reliability of the system continued to amaze. During the Apollo missions—which were really long by the standards of previous flight programs—one-computer operations became the norm. Two computers would be on line for a few hours before and after a powered maneuver.

Antennas at the Deep Space Network station Goldstone, California.

Technicians made a backup tape every hour and a half. That way, if a machine failed while running by itself, its partner could be up and running in a few minutes, with relatively fresh data.

On to the Shuttle

In the 1970s, the Real Time Computing Center did as most other big data processing centers had done: it migrated to IBM 370s. These were virtual-memory machines that used disk drives for secondary storage instead of magnetic tape. Gone were the auxiliary memories. Five of these computers were now in the renamed Shuttle Data Processing Complex. Three would be used during a mission: the control machine, its dynamic standby, and a Payload Operations Control Computer (needed due to the increasing automation of payloads).

Software for Mission Control

In the early days of developing software for the new mission-control system, the IBM engineers were in a building a mile or so from NASA Building 30, where the computers still resided. Software then was still a batch operation. A courier would pick up decks of punched cards from IBM, and zip over to the Control Center. Later, printouts would be returned to the programmers. They got maybe one run a day. NASA finally asked IBM to study the use of TSO, IBM's time-sharing system, for code development. The system boosted productivity; after 1976, Shuttle software development for the ground system would be done on terminals linked directly to the computers.

Even with later attempts to increase productivity and cut down the number of workers (just like at the Kennedy Space Center), there are still a hundred or so software engineers in the Mission Control business today. The Center has 40 display units, and well over 5,000 event lights, that are under direct computer control. It has about 600,000 lines of code in its computers for flight support. During any Shuttle flight, a wealth of data is recorded, analyzed, and ready for use by the controllers to help the astronauts.

Controlling Robots in Deep Space

There is one big difference between mission control for an unmanned spacecraft and for a manned spacecraft. A manned spacecraft has a highly trained crew of experts on board, who can monitor systems closely and handle emergencies. An unmanned space probe is an instance of a rare species: the autonomous robot. It actually spends most of its working life alone, out of contact with Earth. It must be able to handle emergencies with whatever artificial intelligence it can carry along. Sometimes the robot's intelligence is inadequate; witness the recent losses of the Soviet *Phobos* I and II probes, and the U.S. *Mars Observer*.

Refining Mission Control for the Shuttle
NASA's careful and selective improvements have continued. In the late 1980s, the Mission Control computers were upgraded once again, to IBM 3083s. More recently, the black-and-white monitors in the Mission Control consoles were supplemented with color displays on UNIX-based MassComp workstations. As do many workstation networks today, the one at Mission Control uses TCP/IP (a standard network protocol) to link its stations. Individual controllers have brought in even more computers: laptops for keeping required notes and making mission summaries.

All unmanned probes carry some sort of on-board sequencer or computer that will enable them to complete their mission in case the radio receiver fails. They also carry other *failure recovery* capabilities. Mission control for unmanned spacecraft centers on tasks of sending and receiving. The controllers send updated command sequences and navigation calculations; they receive and analyze scientific data and images from the spacecraft.

Computers actually have two major roles in controlling unmanned spacecraft. The first is receiving data and sending commands; the second is helping to analyze the data (which is sent down shortly after it is gathered). To communicate with space probes headed beyond the Earth's gravity, NASA uses just three ground stations in the Deep Space Network. One is located in Goldstone, California, one in Madrid, Spain, and one in Canberra, Australia. Each has an 80-meter-diameter antenna.

The spacecraft are so far away that at least one of the stations is capable of being in contact at any one time. There is, however, usually more than one probe to communicate with; the spacecraft share antenna time on the network. In the early 1990s, for instance, the Network serviced both Voyagers, as well as *Galileo, Magellan, Venus Mapper, Mars Observer,* and *Ulysses.*

Computers at the Jet Propulsion Laboratory

Very soon after the Deep Space Network started up, the Jet Propulsion Laboratory (JPL) bought computers to help format the data coming down from space.

Although it would have responsibility for the Deep Space Network and many interplanetary probes, JPL was not (and is not) a NASA facility. As an independent, federally-funded research center operated by the California Institute of Technology, JPL needed its own computers. The first ones the JPL team used were Scientific Data Systems SDS 910 and 920 machines. The 910 gathered data; the 920 processed it. (These same computers would still be in use in 1985, about a quarter century after they were purchased. They were replaced by more modern mini-computers.)

In the early days, magnetic tapes with output from the 920 would be taken to JPL, and the data analyzed on the institution's IBM 704 (similar to the computer that had been used for Vanguard). Soon there was a sophisticated data network, itself controlled by computers, to move information around.

The Central Switchhouse

JPL has a campus-like site, many buildings interspersed with Southern California palms. Behind the central administration building is the Space Flight Operations Facility that houses the heart of the Deep Space Network.

There have been several generations of computer systems in the Operations Facility over the last 30 years; they have mirrored the evolution of computer use in commercial data processing. After all, the seemingly-endless stream of bits from a space probe is basically a sequence of numbers, like stock quotations or payroll calculations.

In the early 1960s, the Facility housed dual IBM 7094s. Each had a minicomputer-like IBM 7040 as a *front end* (the part of a system intended for interaction with the user). The 7040 would receive and route all sorts of data—either to the project-control centers or to the 7094 for processing. An IBM 1301 disk drive—all 54 megabytes of it—sat "between" the two computers. They both used it as a kind of "staging area" for the temporary storage of data files; it functioned much like what we might call a *file server* today.

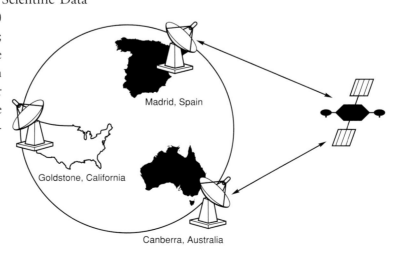

Madrid, Spain

Goldstone, California

Canberra, Australia

The three Deep Space Network stations are placed to provide continuous coverage in any direction.

Engineering duplicate of the Soviet Phobos probe, on display with the visiting Soviet Space Exhibit, St. Louis Science Center, 1992.

Typing Errors Can Be Fatal to Spacecraft

The process of preparing commands for unmanned spacecraft is often tedious and error-prone. JPL created computer-based methods of commanding and uplinking commands to spacecraft. The Soviets did the same, but got caught in an unusual keystroking error during the *Phobos* Mars mission.

The ground station that had been used to control deep space missions was traded for a newer one nearer Moscow. A controller in the new station, using new software, entered a series of commands. Apparently he made a typing error that misplaced a *delimiter* (a piece of code that separates the parts of a command). Parts of two commands "ran together," resulting in something unwanted. Somehow this error survived cross-checking, and the garbled command went up to the spacecraft. When executed, the command caused the on-board computer's flow of control to jump to the wrong location—where there was a safety routine used in ground tests. The routine locked out the attitude control jets, keeping them from firing. The *Phobos* gradually drifted in attitude, and lost its antenna lock on the Earth; since the antenna no longer pointed at the ground station, new commands could not be uplinked to fix things. *Phobos* was never heard from again.

Scratch one expensive spacecraft.

The 7094s did telemetry analysis and predictive work for the ground stations in the network. Navigation was (and still is) done in batch mode by big institutional computers. (These were UNIVACs in the era of the IBM 7094.)

The Deep Space Network was able to take advantage of a double windfall in late 1969. Johnson Space Center had a surplus IBM 360/75, and shipped it to JPL. So began the upgrade of the computers in the Space Flight Operations Facility. By that time, the computer center had grown to a kludge; even its front-end computers had front-end computers. When a second 360 showed up (available because NASA's Electronics Research Center had been shut down), the reincarnation began in earnest.

JPL's windfall 360s did need some work. They were configured at only about half the memory capacity of a similar machine that the Real Time Computing Center used in support of Apollo missions. Worse than that, they ran the RTOS operating system, which did not fit JPL's needs at all. It had the "feature" of using disk storage in a completely different way: as read-only secondary memory. All the read/write data was kept in core memory. Since JPL was supporting multiple missions, and had to transfer data at high rates of speed, they had the operating system changed to match these needs more closely. The new operating system was called (what else?) JPL/OS.

Although everyone thought the bigger machines would be enough to handle the work (hence no front-end computers), that assessment quickly changed. First they connected the 360s to the UNIVAC 1108 that did the navigation calculations, so it would have direct access to the latest data. Then they routed incoming data from specific missions first through a mini like a UNIVAC 1230; since it had already served as a test machine during spacecraft checkout, it could "talk" the spacecraft's language. The 360 was then freed to do the analysis, predicting, and commanding calculations; the 1108 stayed on top of the navigation, and a 360/44 did nothing but image processing.

Even with all JPL's computer power, over 700 human controllers helped in the Viking Mars mission. There were actually four spacecraft—two landers and orbiters. The system was stretched to its limit. The number of mission controllers was twice that of any manned operation.

With the proliferation of spacecraft to work with, and the increase in computing power on board the spacecraft themselves, it became obvious that a new approach to mission-control computing would be needed. Instead of using monolithic mainframes, a distributed system seemed more reasonable. It could work; at about this same time, the engineers at Kennedy Space Center had designed and installed the Shuttle Launch Processing System, their first foray into distributed computing.

Like Kennedy, JPL procured ModComp 16-bit minicomputers. They were ready for

use in 1976, and the new approach was underway. Each major mission would have a mini assigned to it for telemetry processing. A single mini would be shared for commanding. A routing computer in the Space Flight Operations Facility would ensure that data coming in on the worldwide network would be sent to the correct machine. The process for commands would be reversed; the computers would route the uplink words to the correct ground station. Each project would have its own control center, usually outside the Facility. That way a particular space probe could be given further support—by specialized computers—as needed.

By the 1980s, Ethernet (now a standard mechanism for interconnecting computers) had become the backbone of the distributed system, with over 100 computers involved. By the time STS-34 launched the *Galileo* probe, commercial operating systems were in use. Advanced languages such as C and Pascal made up the software. The engineers working in the Deep Space Network—and at JPL in general—had emerged among the space program's most sophisticated software developers. *Galileo* carries the hardware of 1989 on its long mission to the outer solar system, but the efforts of its controllers continue to become more advanced and effective.

The Viking Orbiter and Lander spacecraft.

Capsule Commanders

5

The astronaut in any space mission does not do a lot of piloting in the early part of a flight, while the big rocket boosters are still firing. There is too much happening too fast—all around the spacecraft and launch vehicle—for a human being to be of help. Mostly the astronauts watch for red lights and listen for warnings that might mean a premature exit from their ship and a quick end to the mission. The astronauts are the ultimate backup system; a major failure might disable the automatic abort devices, in which case the astronauts would have to activate them manually.

In the early manned space flight programs, "backup system" would also describe the role of the astronaut in the rest of the mission. The one-man Mercury capsule and its Russian contemporary were built to be fully automatic. There is little evidence that the first man in space, the Soviet Union's Yuri Gargarin, did anything but enjoy the ride. The first woman in space, Valentina Tereshkova, was a parachutist, not a flier.

For the first team of American astronauts, all test pilots, this level of automation was offensive. Their brethren test pilots derisively called them "Spam in a can." They wanted to be in charge of the spacecraft and the mission. They wanted to be pilots in space.

Ed White uses a hand-held maneuvering unit on the Gemini IV flight.

Maintaining a Good Attitude

Since the astronaut feels no pull of gravity when floating in orbit, the terms "down," "up," and so on are difficult to define. The way the spacecraft is pointed relative to some external reference is its *attitude*. For instance, if the horizon is the reference, then "nose-down attitude" means the front of the spacecraft is pointed toward the Earth and away from the sky.

One way a spacecraft changes its attitude is by using small rockets called *reaction control jets*. These jets are easily started and stopped, and use liquid fuel. Conserving enough fuel to control the craft during retrofire and the early part of re-entry is a big responsibility for the crew.

They managed to get control of some parts of the Mercury missions, and with the addition of computers to all later United States manned spacecraft, their role expanded even more. Even though some astronauts would later say, "We're here to serve the computers instead of the computers being here to serve us," they could not have flown their complex spacecraft—on equally complex space flights—without the aid of computers.

No "Spam in a Can"

A Mercury capsule offered few chances to exercise command-pilot privileges. Launched into space atop a military rocket, it was a ballistic object, like a bullet or artillery shell. It depended fully on the booster for guidance into orbit. The booster did not even carry its own computer. A Burroughs mainframe computer at the launch site in Florida received tracking signals from Bermuda and Cape Canaveral, and calculated steering commands that had to be sent up by radio.

The first words John Glenn heard, after the engines stopped firing in the first American orbital mission, were: "John, you are 'go' for seven orbits." That meant the Atlas booster had done its job: since the flight was planned for only three orbits, the capsule had enough speed and altitude to stay up for the entire mission.

All the astronaut could do while floating in orbit was to point the capsule in different directions (change its *attitude*), and use the periscope and window to look around. He did this by moving a small control stick (remarkably similar to those on many video game computers now). Moving the stick sent a signal to a device that would fire small thrusters, moving the capsule in the direction opposite the thrust. The astronaut could flip end for end, or roll, or both.

The most important use of these attitude controls was to make certain the blunt end of the capsule faced the direction of flight when coming back to Earth. Ablative materials covered the wide base of the capsule; these ignited from friction, and burned off at a specific rate during the passage back into the atmosphere, protecting the astronaut. (The narrow end of the capsule had thin metal walls that would have disintegrated from overheating after hitting the atmosphere.) Strapped to the center of the heat shield on the blunt end were three small retrorockets. These had to be used to reduce the speed of the capsule enough so that it would fall out of orbit.

The astronaut got the job of pointing the Mercury spacecraft and pressing the button that fired the rockets. It was important that this be done on time, and with the capsule in the correct attitude; otherwise someone else would have been Ohio's senator. No serious problems arose from giving this much control to the Mercury astronauts, though Scott Carpenter was late firing the retrorockets on the second orbital flight. He landed

John Glenn lifts off to become the first U.S. astronaut to orbit the Earth.

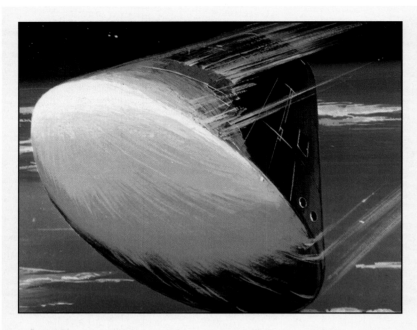

Returning to Earth in a Blaze of Glory

All the non-reusable U.S. spacecraft—Mercury, Gemini, and Apollo—used a "heat shield" made up of *ablative* material to protect the crew during descent through the atmosphere. The friction caused by a rapid passage through even the thin gas of the upper air would quickly cook the surface of the capsule. The basic principle is that heat is shed from the re-entering spacecraft by letting part of the heat shield burn off; as the hot pieces fly away, they take their heat with them.

An Apollo capsule makes a fiery re-entry with its heat shield pointed forward.

dozens of miles past the planned point, and had to spend some time floating in his life raft fishing while the recovery teams frantically searched the ocean for him. At least it was a "human" mistake, and it showed the men were in charge, not the machines.

Twin Pilots Fly the Gemini

Shortly after the young NASA started the Mercury program, the Apollo project to land men on the moon began. The technology gap between the two was tremendous. New, huge booster rockets needed to be developed. Even if they had been ready, the problem of navigating seemed to become more difficult as NASA considered and abandoned various schemes to reach the moon.

At first, the agency was going to try *direct ascent* and return. This was the method most preferred by science fiction movies: launch a big rocket from Earth, and plot a direct course to intercept the

moon's orbit just as the moon reached that point. The lunar lander would have to have enough fuel to slow itself down for a soft landing, and then launch back to Earth with the astronauts' rock collection. The problem was that the largest rocket then under serious development, the Saturn V, was only powerful enough for a one-way trip via direct ascent. Needless to say, the sort of people who would volunteer for such a mission were hardly the type NASA wanted representing the United States!

Even though an even-more-powerful launch vehicle, the Nova, was on the drawing boards, it could not be expected to fly before the Soviet Union would be ready with a lunar mission of its own. In the days of the 1960s Space Race, waiting for Nova was unthinkable.

NASA studied several alternatives to direct ascent. The Saturn used most of its fuel climbing out of Earth's *gravity well* to an altitude where it could go

Launch of the Saturn V, the largest booster ever built in the U.S.

The Law of Gravity and Rocket Fuel Consumption

Launching a rocket from the surface of the Earth means the force of gravity is highest at the beginning of flight, and steadily decreases as the vehicle climbs out from what is often called the "gravity well." Flying from the Earth to the moon and returning means that large amounts of fuel are needed to get far enough out, to the point where the moon's gravity exceeds the strength of the Earth's gravity. From there, no fuel is needed to get *to* the moon, though stopping is another problem. Things are reversed in the other direction: a relatively small expenditure of fuel helps to escape the gravity well of the moon, then a free ride to Earth is followed by a bone-crunching slowdown in the atmosphere.

Apollo Lunar Module's ascent stage climbs from the moon toward the Command Module. Only a small rocket motor was needed to overcome the moon's gravity, which is one-sixth Earth's.

The in-flight linkup of the Apollo spacecraft on the way to the moon.

into orbit. Therefore engineers studied various forms of rendezvous in low Earth orbit. In one scenario, a tanker could refuel the manned lunar vehicle prior to its departure for the moon. Once the problems of passing super-low-temperature fuels from one spacecraft to another came up, however, these schemes were quickly abandoned.

Eventually, engineers at Langley Research Center in Virginia suggested a different scheme: the specialized modules of a two-section spacecraft would rendezvous in lunar orbit instead of Earth orbit. The first two stages of a Saturn V, and some of the fuel in the third stage, would lift the lunar landing vehicle and a *Command Module* (in which the astronauts would travel) into Earth orbit. After

making about one and a half trips around the globe, the third stage of the Saturn would fire again, and inject the Apollo vehicles into a flight path for the moon. Shortly after the third stage burned out, the astronauts would turn the Command Module (and its attached *Service Module*) around, and hook the nose into the docking collar of the lunar landing craft.

Connected together, the space vehicles would fly for two and a half days to the moon. Partly captured by the moon's gravity, they would swing around to the far side. At a critical point in their trajectory, the engine on the Service Module would fire, slowing them down enough to drop them into an orbit around the moon. Two astronauts would then crawl from the Command

Module to the Lunar Module, check out its systems, and disconnect for the descent to the surface. Explorations complete, the two astronauts would lift off from the lunar surface in the top part of the landing vehicle, leaving the bottom part as a monument to human exploration. This small *ascent stage* would then find and dock with the Command Module.

Once the rock samples, other materials, and the astronauts were safely in the Command Module, the ascent stage could be disconnected and sent hurtling to a crash on the moon. The three astronauts, together again, would fire the Service Module's engine to leave the moon and start the trip back to Earth. After another two and one-half days, they would separate the Command and Service Modules. The Command Module would have an ablative heat shield like the one on Mercury. After a ride into the atmosphere, three

parachutes would open, and let the Command Module down softly into the ocean.

What's in a Word?

A computer *word* is usually the number of bits contained in a register in the memory. These bits are either subdivided or attached to other groups of bits to make different sorts of "words." Many *instruction words* have a few bits set aside for instructions, and the remaining bits for the address of the data on which the instruction will operate. For instance, Gemini used four bits for instructions and nine bits for addresses. *Data words* can consist of ever-longer collections of words; Gemini's data words were 26 bytes long.

Using this flight profile meant some difficult problems had to be solved. For the larger maneuvers, such as leaving Earth orbit and calculating the best time to fire the Service Module's engine to enter lunar orbit, ground-based computers could be used to help navigate. But Earth-based radars and computers were out of the question to help guide the lunar landing. It was just too far for accurate ranging; even at the speed of light, a command signal would take over a second to arrive. A lot could happen to a falling spacecraft in a second.

NASA decided very quickly that the most difficult parts of the lunar mission ought to be tried in Earth orbit first: rendezvous with another

The Apollo Command and Service Modules in lunar orbit. Note the restartable main engine at the far end, and the conical Command Module, with its docking collar, at the near end.

The Lunar Module of Apollo XIV. The descent stage acts as a launching pad, remaining on the moon when the astronauts use the ascent stage to return to lunar orbit and their rendezvous with the Command and Service Modules.

spacecraft, long periods of weightlessness, and automatic entry into the atmosphere. At the time, no two spacecraft had even come close to each other, let alone flown in formation. The longest anyone had been weightless was a few hours, and space medicine experts still feared the worst.

The need for automatic entry came from worries that the astronauts would be too tired to fly the Command Module manually into the tiny "window" in the sky they must hit for a safe return to Earth. The angle had to be nearly perfect; approaching Earth from the moon at 25,000 miles per hour (50 percent faster than orbital speed) would pose two dangers. If the command ship came in at too shallow an angle, it would skip off the upper atmosphere like a stone across water, and go into deep space with a crew doomed to die from lack of oxygen. If it came in at too steep an angle, it would become a meteor. NASA wanted something to guide the astronauts that would not suffer from fatigue: a computer.

The Gemini space flight program aimed to solve these problems. It would fly with a crew of two for up to two weeks in low Earth orbit. It had a docking device to hook it to other spacecraft. Most importantly, it had a computer to assist in rendezvous and in automatic landings.

The Gemini Computer

NASA chose IBM to build its first space computer. Unfortunately (as with many aerospace projects since), the spacecraft was designed before the computer. By the time the IBM engineers got their marching orders, the bay left for the computer was 19 inches by 15 inches by 13 inches—with a weight limit of about 60 pounds. Nowadays that would seem a huge allotment of space and weight for a computer. In 1962, however, the average computer required a pristine, air-conditioned room, and enough floor space and energy to house a large family. For the technology of the day, the available space was tiny.

The main difference between the Gemini computer and those on our desks is the *integrated circuit*, which we know as the computer chip. Many thousands of tiny transistors, sometimes millions, can now populate each chip. In spaces that housed one or two simple parts of the Gemini computer, we can now pack devices as complex as entire computers. But computer chips were not available in 1962. The Gemini computer used discrete components, and had a *core memory* for the storage of programs and data.

Core memories are nearly extinct in civilian applications, and are becoming so in the military. In 1962 they represented the state of the art. A core memory stored individual bits of information on ferrite (iron) rings so tiny that 25 of them laid side-by-side would cross a dime. The ones and zeroes of the binary numbers were represented by polarizing a magnetic field one way or the other on the rings. Each ring had three wires running through it; two were used to change the ring's

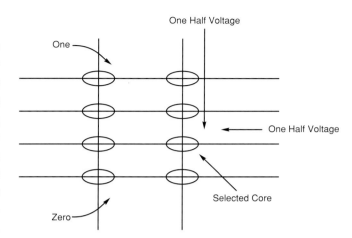

Schematic of a single core-memory plane.

polarization, and the third sensed the change. The rings themselves were strung in rows and columns that formed planes; the number of planes determined the size of the computer word.

Getting the Right Address

An *address* in a simple computer memory leads the software to a specific location. Most addresses are numbered in order. For instance, memory location 1336 in base ten means "the 1,336th storage place for data or program instructions." Complicated memories (like that on Gemini) may require a portion of the address for the sector, and the remainder of the address for the location *within* a sector.

Let's say we wanted to store the binary number 11001100 in a core memory of the Gemini era. The memory would have eight planes, consisting of as many rows and columns of wires as needed for the entire storage. We would start by placing a "1" in the plane that stores the first digit of the

Gemini spacecraft in orbit. The flight computer is located between the pressure wall and the skin of the spacecraft, under the end of the words "United States."

number at a specific location (which will be the same in all the other planes, for all the other digits). Assuming the cores all start polarized as zeroes, we would need to split the voltage that is required to "flip" the rings' polarization; we'd send half down the "row" wire, and half down the "column" wire. At the intersection of the two wires would be the core we wanted to use. Only at this intersection would there be enough voltage available to switch the polarization. All other cores on the rows and columns would have received only half the voltage.

The remaining "1" bits of our binary number would be stored the same way, each in a different plane of rings. The zeroes would not need to be changed. When the time came to read the information in the cores, the voltages would be sent again to all the cores used to store the data word. As the cores switched polarization, the sensing wires running through them would sense the change; this information would go to the memory channel that requested the data. Basic core memory was "destructive": reading the data destroyed it by removing it from the cores that contained it. The data could be restored, however, simply by sending two voltage pulses of the same intensity used to store the data originally.

The use of discrete components and core memory made the Gemini computer a snail. It took nearly a half-second to do a multiplication, and twice that for a division. (Modern desktop workstations can do millions of either computation in a second.) The relatively "big" components took longer for electrical pulses to cross.

The Gemini computer had one of the smallest *instruction sets* of any computer. Now, small instruction sets are not necessarily bad—many of the newest PC workstations achieve greater speed by using smaller sets of internal instructions to guide the processing. But these "reduced instruction set computers" (or *RISC architectures*) have an advantage unavailable to Gemini: their instructions are vastly more powerful than those in older computers. Thus fewer instructions are needed to accomplish some of the same processing—but even a RISC machine has hundreds. In a Gemini computer word, only four bits could be used for instructions. The largest decimal number that can be represented in four bits is 16, so there were

Schematic of the core memory of the Gemini flight computer.

16 instructions. Clearly, not much of the computing power available at the time was portable enough to go into space.

Perhaps more surprising is the unusual memory arrangement in the Gemini computer. A computer word is used for either instructions or data. Nearly all current computers use some multiple of the eight-bit byte as their word size: we are accustomed to 8-, 16-, 32-, or 64-bit machines. The Gemini computer had a 13-bit instruction word, and a 26-bit data word.

Scaling

In a computer using fixed-point arithmetic, the programmers are responsible for "scaling" data to ensure its accuracy. To do this, the engineers have to use machine-language instructions that shift bits either left or right of the decimal point.

Part of the reason for this unusual word size was the need to fit the machine into a limited space (which its discrete components rapidly filled up). We have already seen that the first four bits of the instruction word contained the code that actually operated a Gemini computer. That left nine bits for *memory addressing*.

The Gemini computer engineers had a dilemma: they wanted a machine that was simple and light enough to fly on the mission, but also accurate enough to do crucial calculations. To represent the most accurate numbers possible, they had to make the data word 26 bits long. This was because the machine was a "fixed-point" instead of "floating-point" design.

In a *floating-point* machine (like nearly all desktop PCs), numbers are represented using several digits and a movable decimal point. This is much like scientific notation, where numbers take the form of a decimal multiplied by a power of ten: "1.634 times 10 to the eighth," for example, or "50.3

A Gemini spacecraft on display at the St. Louis Science Center during the International Year of Space, 1992. Note the absence of the aft skirt (discarded prior to re-entry).

The Gemini computer core memory had just over four thousand 39-bit locations. The Gemini designers termed these locations "words," and subdivided them into "syllables." Each 39-bit "word" had three of these (13 bits each). Any of the 13-bit syllables could be used to store one of the 13-bit instructions. However, a *26*-bit *data word* could only be stored in the first *two* syllables, and could never be in the third syllable. Got that?

Worse yet, you would think that if a 39-bit word had three syllables of instructions, the computer would execute them one after another, finish with the word, and then move on to the next word in line. Not so! The memory was also divided into 18 sectors. To execute a program, the computer would use all the instructions in the *first* syllables of the words in a sector, then all the instructions in the *second* syllables, then all the *third* syllables, and then move on to the next sector.

Neil Armstrong and David Scott ride into space atop Gemini VIII's Titan II launch vehicle.

times five to the minus ninth." In a *fixed-point* machine, however, the decimal point is always in the same place in a data word, and the actual data must be "scaled" around that point. Where you place the fixed decimal point will put a limit on how many numbers you can put to the left or right of the decimal, which in turn can limit the variety and accuracy of the possible calculations. This type of computer is much easier to design and build, but it gives the programmers fits. And no memory design gave a programmer more trouble than the Gemini memory.

You can probably see the next problem coming: the 26-bit data words extended into the second syllable of the 39-bit words. This meant any instruction stream using second syllables would have to be programmed to *jump over* any data it might find. Otherwise the hapless data would be executed as an instruction. Whew!

This mess might not have been so bad, had there been a language that could be used to simplify programming the Gemini computer. These days we use *high-level languages* (such as BASIC or C) to construct programs. They hide machine-specific details such as those just described for the Gemini computer. The programs written in such languages are passed through a *compiler* program, whose output is machine instructions. But no compilers existed for a special-purpose, custom-built computer such as the Gemini. The programmers had to write instructions for the computer in *octal* (base-eight) numbers, by hand, and maintain them the same way.

There is a saying in the computer business that "software expands to fill the memory allocated for it." When the IBM engineers first studied the requirements for the Gemini computer system,

they figured the 4,096-word memory would be more than enough. But as the Gemini missions proceeded, more and more things moved from the "wish" list to the "must-have" list, and the software became too large for the memory to hold.

We commonly have this problem even today. All the programs you might want to run on your PC rarely can be stored in its main memory at the same time. You have to move them in and out of main memory as you need them.

Nowadays the most common secondary storage is the *magnetic disk*—either a "hard" disk permanently installed in your machine, or a "floppy" that you can carry around with you. In 1962, magnetic disks were the size of a huge serving platter, and required large amounts of electric power.

The Agena upper stage launched by an Atlas booster.

Computers in Space

They were completely unsuitable for spacecraft; any rocket or airplane designer must consider size, power requirements, and weight. Even had there been a rocket powerful enough to heft the disk and its power source along with the payload, a big spinning plate inside a floating spacecraft would have acted as a gyroscope!

The most common form of secondary storage at the time of the Gemini program was magnetic tape. Several small data-tape systems had been built for satellites such as the Orbiting Astronomical Observatory. NASA bought a similar one for Gemini that expanded the storage capacity for the computer over seven times.

Unfortunately, there was no room left in the capsule for the tape unit. It was installed in the aft skirt that contained some maneuvering engines, and discarded before re-entry into the atmosphere. Therefore it was very important that the astronauts load the right programs before they threw away the tape unit! When the Gemini rocket took off, the computer contained the programs to back up the Titan booster's guidance computer and the orbital insertion program. Other programs, such as the one to control the return to Earth, had to be loaded later—from the tape unit—as they were needed.

NASA worried about the reliability of all systems on manned spacecraft, but it seemed they worried the most about the computer systems. Tape memories in 1960s computers had data-transfer error rates of about one bad bit in every 100,000. Not bad, but NASA and IBM wanted *one bit in one billion!*

The Gemini flight computer (inset). Note the odd shape dictated by its location within the spacecraft.

They achieved this reduction of error by triple-recording every program, and then passing all three versions bit-by-bit through a circuit that "voted" on them. An error would occur only in the (extremely unlikely) event that all three recorded versions of the program had the *same* bit in error. The tape memory also had tiny flywheels and weights, like the balancing weights on automobile wheels, installed on its motors and reels to keep them steady while transferring data.

The very first time the tape unit flew into space on Gemini VIII, all this concern for reliability got its first big test. Neil Armstrong and his crewmate David Scott completed docking with an Agena upper-stage rocket that had been put into orbit first. Just under half an hour after the capsule mated with the Agena, an attitude-control thruster failed and sent the spacecraft spinning. Armstrong and Scott had to use the re-entry control system to stop the spinning. Then, low on fuel, they had to load the re-entry program from the tape and come right back to Earth. The software transferred successfully; the astronauts returned safely.

"Flying" the Gemini Computer

We saw why the Gemini spacecraft had to have a computer: to rehearse critical Apollo mission phases such as rendezvous with other spacecraft and automatic re-entry. In the end, the computer had six programs: Executor, Prelaunch, Ascent, Catch-up, Rendezvous, and Re-entry.

The "Executor" program was actually a simple operating system—much simpler than MS-DOS or UNIX—that selected and executed the appropriate program for the current mission phase. "Prelaunch" helped the ground crews check out the functionality of the computer and other parts of the guidance system before committing to flight (see Chapter 2). "Ascent" received the same data as the guidance computer in the Titan launch

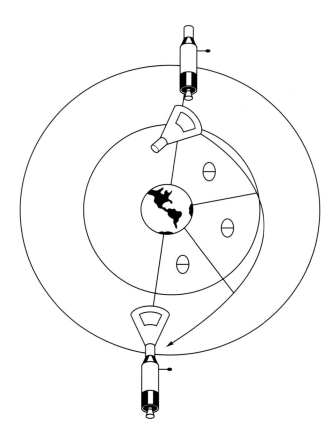

The docking maneuvers between Gemini and Agena were practice runs for later docking maneuvers on lunar missions.

vehicle. If that computer failed, the spacecraft computer would take over and fly the rocket. "Catch-up" was the first of the two rendezvous programs. It enabled the crew to change orbits to intercept the Agena upper stage or another spacecraft. The final closing and station-keeping maneuvers were under the control of the "Rendezvous" program. Finally, "Re-entry" pointed the spacecraft in the correct direction for retrofire, fired the engines, and maintained attitude from 400,000 down to 90,000 feet.

The Gemini computer had three *interfaces* with the crew: some control switches, the *Manual Data Insertion Unit* (MDIU), and the *Incremental Velocity Indicator* (IVI). The control switches were few and simple: a malfunction light, a light that blinked when the computer was calculating, a

Gemini/Agena docking sequence during the Gemini VIII mission: The Gemini spacecraft finds Agena.

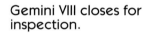

Gemini VIII closes for inspection.

Gemini VIII approaches nose-first to complete the rendezvous.

mode switch that helped choose the programs, a start button, and a reset switch.

Spacecraft have very limited power supplies, so the astronauts turned off the computer unless it was needed right then. If you pull the plug on your home computer while a program is loaded into memory, the computer "forgets" the program, and you have to reload it when the power comes back on. The core memories of Gemini, however, did not need electrical power to "remember" programs and data. Since they consisted of polarized iron rings, like magnets, the data stayed put. When the astronauts powered up the computer, it took 20 seconds to become ready; then, after they entered any new data, they could press the start button to send the program running. The reset switch was like the red button on many PCs that allows the computer to "reboot" itself without turning off the power.

To understand how to use the Manual Data Insertion Unit (MDIU) and the Incremental Velocity Indicator (IVI), let's fly the Gemini to a rendezvous with an Agena. The Agena upper stage was a rarity in the early 1960s: it had a rocket motor that could be turned on and off. (Most rockets of the time burned continuously until their fuel ran out, or until they reached some predetermined speed. Even if they had fuel left, they were not designed to start again.) The Agena could be carried into orbit using the same Atlas booster that had lofted the Mercury capsule.

Once there, it awaited a Gemini spacecraft to push its nose into the docking collar on the Agena. After that, the Gemini had the use of the Agena's engine to change orbits and conduct flight experiments. The Atlas-Agena would be launched first. About 100 minutes later, the Gemini would follow it

up. Once both were in established orbits, the chase to mating would begin.

The first thing the crew needed to do was enter any special parameters into the computer memory. As an example, one of these parameters was the angle of the orbit the spacecraft would traverse before another firing maneuver would be calculated and executed. To rendezvous with the Agena, the astronauts had to change orbits to intercept the Agena's orbit, preferably when the Agena would be there.

If you watch *Star Trek*, you know it is only necessary to point the nose and fire the engines. Unfortunately, the Gemini's maneuvering engines had nowhere near the thrust and fuel of the fictional *Enterprise*. They had to make small changes in the orbit, each change bringing them closer. It was not unusual for a rendezvous to take nearly six hours, and require nine or ten associated maneuvers. One way of specifying when to make the next maneuver was to enter an angle into a specific memory location; the next set of calculations would start when the spacecraft reached the corresponding point in the orbit. If the angle were, say, 30 degrees, then the astronauts needed to load "30

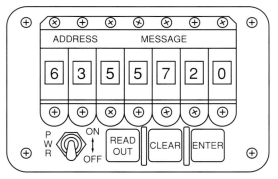

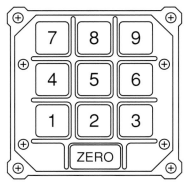

The Manual Data Insertion Unit (MDIU).

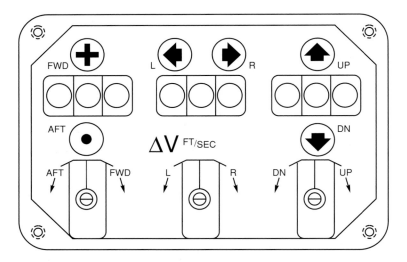

The Incremental Velocity Indicator (IVI).

degrees" into memory location 93 of the Gemini computer. To do this, the pilot used the MDIU.

The Manual Data Insertion Unit itself had two parts: a display and a keypad. The keypad was a simple zero-through-nine device. The display had seven digits. If you look at the figure that shows the MDIU, you might think the seven digits look more like the numbers on your car's odometer (assuming it is not digital) than the seven-segment numbers on most calculator readouts. This is because they *are* like the numbers on your car's odometer; the MDIU used the same principle: stamped numbers on mechanical wheels. The first two digits were the memory location, the next five the number. This meant there were 100 logical memory locations accessible by the crew (numbered 00 to 99). The five "message" digits could be reduced to four by a negative number; all negatives had a "9" in the first digit place.

Below its number display, the MDIU had a power switch and three push buttons. The power switch was just that; the crew flipped it to "on" when they needed the computer. Pushing the "read out" button would display (in the five "message" digits) the data stored in the location you specified in the first two digits. If you wanted to see what was in mem-

ory location 63, you would press "6," then "3" on the keypad, then "read out." The cylinders of the five-digit message would rotate until the number showed up, much like a Las Vegas slot machine.

If you wanted to put something in—for example, specify a maneuver by entering "30 degrees" into memory location 93—you would press "9," then "3," then "00300" (due to the scaling involved), then "enter," and the data would go to the right place. The "clear" key enabled the astronauts to correct mistakes they made. Once all the data needed for the orbit change was in the computer, the astronauts pressed "start," and the computer was off to the races. The result of its first computation appeared on the IVI.

Three Reference Points That Stay in Place

An aircraft or a spacecraft moves in three axes: pitch, roll, and yaw. *Pitch* is movement up and down along the longitudinal axis (the axis down the length of the craft; therefore, nose-up and nose-down). *Roll* is movement around the longitudinal axis. *Yaw* is side-to-side movement of the longitudinal axis. (A diagram in Chapter 7 illustrates all three.)

If you look at the figure of the Incremental Velocity Indicator, you can see it consisted of three three-digit displays: a cross and dot above and below the leftmost display, and around the second and third display, two sets of what looked like '64 Ford Fairlane turn signals.

The numbers showed changes in the spacecraft's velocity in feet per second (as marked on the display), in all three axes of flight needed to maneuver the spacecraft to the next point in its orbit. The cross meant "forward," the dot "rearward," and the arrows either "left/right" or "up/down," depending on the axis in question. For instance, the number "035" in the center of the IVI, with the "left" arrow illuminated, meant the astronauts should add *35 feet per second* to the *left yawing vector*. The cross and dot meant additional velocity forward, or reduced velocity aft.

Once the numbers appeared, the spacecraft commander (in the left seat), would begin to fire the maneuvering thrusters while pointing the spacecraft with the attitude controller. The objective was to achieve the velocity changes dictated by the computer. As the engines fired, the computer would receive data in real time from a device that measured acceleration. That data could be used to calculate updates for the IVI. If he was doing his job right, the commander would see the numbers move toward zeroes on all axes. Even if his maneuvers were not perfect, there was always another firing coming up to let him make corrections.

While the commander flew the spacecraft, the pilot in the right seat used a backup computer to check calculations. The backup in 1965 and 1966? Paper, pencil, and a few pre-calculated charts. The hand-held electronic calculator was not invented yet!

The other main use of the Gemini computer was automatic re-entry. A Gemini spacecraft could effectively change its impact point by up to about 400 miles on its descent track, or forty to fifty miles left or right of it.

Astronaut Ed White on the first U.S. space walk.

On the Gemini IV mission—in which Edward White did the first extravehicular activity on a U.S. space flight—the crew entered all the parameters for an automatic landing, but the computer would not turn off when the switch was thrown. The power was cut off by pulling the circuit breaker, and the landing had to be done manually.

The astronauts missed the landing site by 40 miles. The next landing was worse, and it was not because of bad piloting. Gordon Cooper (a Mercury astronaut still in the space program) and Charles Conrad spent eight days in space before landing 89 miles short of their intended site. Incorrect navigation coordinates had been transmitted to the computer. Later flights averaged landings about three miles from the planned points; Gemini IX splashed down within a mile of the recovery carrier *Wasp*.

Legacy of Gemini

The Gemini program completed all of its major flight objectives. James Lovell and Frank Borman spent two weeks in orbit in Gemini VII, longer than a round trip to the moon. They proved that nothing worse than a shower and a shave would be needed by a lunar astronaut on his return home.

Gemini spacecraft rendezvoused with the Agena, with a spare docking collar after an Atlas-Agena had failed to achieve orbit, and with each other after another such failure. The Gemini spacecraft made several highly accurate automatic re-entries. The techniques needed for the moon landing were tested and proven. Also, Gemini astronauts gained hours of flight experience that would carry over to Apollo.

Note the maneuvering rocket nozzles in the aft skirt of this Gemini spacecraft. They were used for changing orbits and descending to Earth.

Most important, the Gemini computer proved that one of the more fragile electronic devices then known could be made spaceworthy. In 1962, the most common computer failure was a hardware failure. Now it is more likely to be a software failure.

The Gemini computer was a first step in making "flight computers" that could be counted on to control an aircraft or spacecraft in perilous conditions. One of the Gemini computers successfully restarted after two weeks' immersion in salt water. Another went to NASA's Electronics Research Laboratory, to be studied for possible uses as a controller for vertical- and short-takeoff-and-

landing aircraft. Thirty years later, STS-34 would lift off on its *Galileo* deployment mission with *five* general-purpose computers aboard the Shuttle—*each* 150 times faster (and with 80 times more memory) than the Gemini computer.

As the Gemini missions progressed, work on a much more complex computer system went on in Boston. The new system would do most of the work navigating to the moon and controlling a much larger spacecraft.

Frank Borman and Jim Lovell sit in the raft next to their Gemini spacecraft after a then-record 14 days in orbit, 1965.

To the Moon

6

Navigating to the moon ought to be easy. It's 2,000 miles across—a pretty big target to hit. Nevertheless, a quarter of a million miles is a *long* range from which to "take a lead" on a moving target—even such a big one. Even the smallest error in speed or direction magnifies into a big miss. Worse is the return: Earth may be a much bigger target, but hitting the small "window" (ideal re-entry point) in its atmosphere at 25,000 miles per hour—and at a perfect angle—is anything but easy.

Navigation, more than rocket technology, was the problem that set the pace of development in the Apollo program. NASA shot *Ranger* probes at the moon again and again before a "hit" could be scored. It's one thing to send an unmanned spacecraft off into the darkness of interplanetary space—quite another to send three astronauts with no hope of rescue. The guidance and navigation system *had* to work.

Apollo's target.

Born in Boston

In Cambridge, across the Charles River in the Greater Boston of the late 1950s and early 1960s, Charles Stark Draper led the Instrumentation Laboratory at the Massachusetts Institute of Technology. Dr. Draper was a distinguished designer of aerospace control systems. His team at MIT had built the guidance system for the Polaris submarine-launched ballistic missile. The Polaris program was not only a technical marvel, but a management triumph as well, coming in on time and budget. The guidance problem on Polaris was tough: hit a target with a relatively small warhead from a moving launch platform.

Military rocketry had faced such tough problems for centuries. From the first Chinese rocket artillery (in the thirteenth century) to the German V-2, to the U.S. Atlas and Titan, ballistic missile builders had struggled to keep their creations on course. Their common objective was to deliver the payload (in this case, the warhead) as accurately as possible, doing the most damage with the least size and weight.

The problem posed difficulties in balancing the factors of accuracy and effectiveness. Though relatively accurate, the V-2 (for example) was inefficient compared to the bombload of a typical aircraft. From a purely military point of view, it scarcely mattered where in London a V-2 hit; the one-ton warhead could not damage too wide an area. The V-2's greater potency was as a terror weapon that kept hitting London, coming out of the high atmosphere with no warning.

At the other end of the scale were the early Soviet intercontinental ballistic missiles. Grossly inaccurate, they still carried warheads of such destructive power that a miss was almost as effective as a hit.

The early Polaris missiles were small, with relatively small warheads. To do sufficient damage to a target area, they needed to nail the center of it. The guidance system of a Polaris had to be constantly told the current position of the launching submarine, and had to recalculate the missile's trajectory based on the firing location. It needed a computer.

A Trident missile, successor to Polaris, bursts from the water on a test flight.

Paranoia, Delay, and Traffic in Space

In the last chapter, we saw that once the lunar-orbit-rendezvous flight profile became the accepted plan, there were several reasons to have a computer on board the Apollo spacecraft to help in navigation. At the time, NASA also had other reasons; some appear naive in retrospect.

Some people actually thought that if the Russians became sufficiently jealous of U.S. success in space, they might try to disrupt communications with Mission Control. The agency wanted to negate any hindering effect of hostile jamming by giving the astronauts a means of doing their own precise calculations. (The fear turned out to be unfounded.)

NASA also wanted to prepare for later long-duration missions such as flights to Mars (which at that time seemed possible before the century was out). At such great distances, communication delays would become more of a problem. Radio signals could travel at the speed of light, no faster; calculations from Earth would take minutes to arrive. The on-board computer would give the crew a "local" processing capability.

The need proved genuine, though the rationale would later seem premature. Generations would elapse (how many, we don't yet know) between the first lunar landings and the first manned flights to the planets. In the meantime, on-board computers would become ubiquitous features of spacecraft; eventually any space flight, even short ones like Shuttle missions, would be unthinkable without them.

Finally, NASA worried that if multiple missions were in space at the same time, it would saturate the control centers. This concern turned out to be real enough (as we saw in the discussion of control centers for unmanned missions), but it is still not a factor in manned space flight. Manned space traffic is likely to remain sparse for many years.

Yet, in the midst of this sobering military technology, there remained great potential for space applications. Some of Charles Draper's engineers in the Instrumentation Laboratory had their eyes on the stars instead of the cities of potential enemies. (They had some famous predecessors; during World War II, Wernher von Braun had been arrested by the German SS because they thought too much "space" work was coming out of the Peenemünde development center.) At MIT, Raymond Alonso and Eldon Hall proposed designs for aerospace computers in the late 1950s. It was good foundation work for their Laboratory's next big project: a rocket to the moon. From the celestial point of view, a planet was as much a "moving launch platform" as was a submarine. Unlike Polaris, however, this rocket would have to reach a target that would be moving as well.

Legend has it that the Kennedy brothers, Jack and Bobby, frequented a certain oyster house in Boston, and would hold far-ranging discussions on technical and political subjects with an array of guests. Later, as President of the United States, Jack Kennedy would wrestle with the idea of making a national commitment to go to the moon. He and Draper debated the navigation problem. Trying to convince NASA administrators that it could be done (and so sure of success), Draper offered to fly as navigator on the moon mission.

In the end, Kennedy addressed Congress with the challenge to send men to the moon and return them to Earth by the end of the 1960s. This speech came three weeks after Alan Shepard's 15-minute sub-orbital flight, the first of the Mercury program. Such an ambition could not have been built on such a small step. It had to be based on the promises of von Braun's rocket team to build the Saturn, and of Draper's team to guide the spacecraft—first to the moon, and then to Earth.

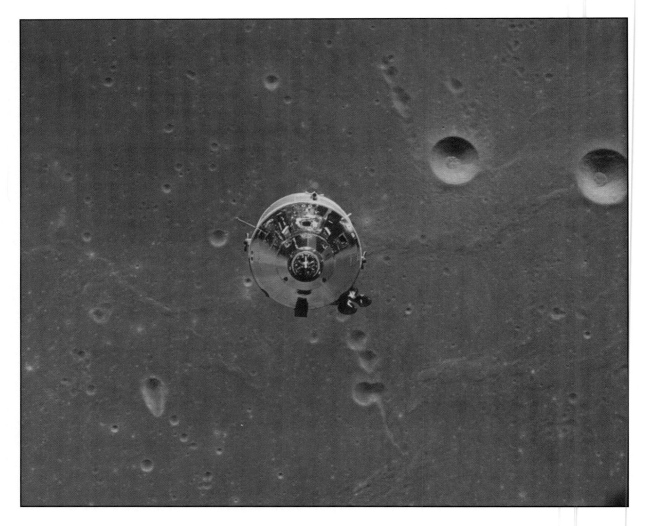

Apollo XI Command and Service Modules above the Sea of Fertility.

Growing a Computer

When MIT started work on the Apollo Guidance Computer in 1961, they put their best and brightest on the project. The Polaris team moved over largely intact. Milt Trageser managed the development, and David Hoag led the technical effort.

Their biggest problem turned out to be determining what was really necessary to build. NASA had given the Instrumentation Lab what amounted to a one-page specification; it revealed little more than the space agency's ignorance of what needed to be done. As development progressed, the MIT engineers finally realized the magnitude of the project—and the extent of the needed innovations. Later they would say that had they known

the real situation in 1961 (the year they started), they would have thought their task impossible with existing technology. Luckily, they would soon get a technology boost.

One decision that came out of the plan to use lunar-orbit rendezvous was to put a computer in both the Command Module and the Lunar Module. The computers would be identical, but the *software* would be tailored to the mission of each spacecraft. The plan exploited the fundamental flexibility of the digital computer: software could make it do different things in different situations. Until that time, hard-wired circuits made up guidance and control systems. They could do only what they were built to do, nothing else. In fact, they could not be used in two different rockets or spacecraft without some redesign and rebuilding. A computer-controlled guidance system needed only to run a different program.

In the NASA world of obscure acronyms, the two-computer system came to be called PGNCS, pronounced "pings" (an acronym made up entirely of consonants couldn't be pronounced anyway). It stood for "Primary Guidance, Navigation, and Control System."

At first, the MIT team intended to use the same design approach as on the Polaris computer. It was a familiar technology—made up of discrete components, just like the Gemini computer. It also had the same sort of destructive memory readout, which was bothersome. It meant the designers would have to include data-rewriting circuits, making make the computer bigger and heavier. The older technology also presented problems in timing and speed. Hoag and his team began to look for newer technology to solve the problems.

The integrated circuit (later to become familiar to computer users as the "chip") had been invented in 1959. It contained several transistors on a single piece of silicon wafer. The Apollo computer designers settled on an integrated circuit of three transistors and four resistors that implemented a logic operation called a *NOR*.

In Boolean logic (the kind used in computers), a NOR inverts the results of an OR operation. Say the three inputs to the chip are "1," "1," and "0." Then the result of the OR is "1," and the output of the integrated circuit is "0." By arranging these chips in various orders with different wiring schemes, the engineers could build *adders*

(complementing circuits needed for subtraction), *comparators*, and all the other components of an *arithmetic logic unit*.

Eventually, the Apollo computer had 5,000 of these NOR gates. This technology was so new that the production demands of the Apollo computer sucked up over half of all chip production in the first couple of years. As the use of integrated circuits skyrocketed in the middle Sixties, the Apollo needs shrank in proportion; by the time the astronauts actually landed on the moon, the chips in the Apollo computer were "ancient" technology.

Memories for the Moon

The MIT team had the usual size, power, and weight restrictions. This time the computer could be designed along with the spacecraft, instead of as an afterthought; the packaging could be a little more rectangular (two feet by a foot by a half-foot) than the funny shape of the Gemini computer. Also, instead of riding exposed to vacuum in the last available nook (as the Gemini computer had done), the Apollo computer came in from the cold. It was inside the pressure wall of the spacecraft so the crew could repair it (the original concept even called for spare computer parts).

The size of the Apollo computer word was 16 bits—still small—since it was a fixed-point machine like its Gemini predecessor. Data words

Early integrated circuit. Though grossly large by today's standards, it has hundreds more logic gates than the chips used on Apollo.

could be made bigger by hooking 16-bit words together. One of the three-dimensional vectors that told spacecraft position took three of these double-length words to store. Also, since the address part of the instruction word was small, eventually there had to be some complicated addressing schemes. (Addressing is discussed in Chapter 5.)

The memory had started very small as well (at least for the purpose of going to the moon and back). The original design had about 4,000 words in permanent storage, and only 256 words in core memory, like the Gemini memory (see Chapter 5). That was not a misprint: *256 words*, not "K-words" (*thousand* words) as in a PC. The only reason for erasable memory was to store data that would change during flight, like some counters and the position vector.

The permanent memory was fixed storage for the programs that ran the spacecraft. The Apollo permanent memory functioned much like the read-only memory in your home computer (which contains the programs needed to "boot" the machine, and other invariable software). NASA expected the astronauts should have no reason to change the permanent memory's programs during flight.

As the true requirements for the lunar mission revealed themselves, the complexity and size of the programs increased. When the Apollo finally went to the moon, each of its two on-board computers had about 36,000 words of non-erasable and 2,000 words of erasable memory.

Cores to the Moon

The erasable Apollo memory was the same type of core used in Gemini. Coincidentally, core memories had originated in another MIT lab a decade earlier. The Whirlwind computer, built by MIT as the basis for new types of Navy flight simulators, used core first. Core memory is resistant to radiation, whereas the average chip is not. This is why it is popular in spacecraft and military aircraft.

For spacecraft, core memory is good because it retains data even with the electricity off. Spacecraft are always short on electricity. The downside of core memory is the volume it fills. Only one bit per core meant that a 38,000-word Apollo memory would have needed over *600,000 cores*, each with sensing and "flipping" wires (see Chapter 5). It would have filled all the space allocated for the computers, and then some.

To compress the space needed for memory, the Apollo designers used *core rope*. This design concept got its name because the cores lined up parallel along hundreds of feet of wiring. Each of the cores in a core rope was permanently charged as a "one" (unlike regular cores that could represent either a one or a zero). Also, unlike the one-bit cores, each Apollo core could store up to *64 bits*! That was a significant increase in density.

The permanent memory was the only part of the Apollo memory to use the "core rope" approach. To see how it worked, look at the second core from the right in the diagram of the Apollo core rope memory. Let's say we wanted to store "0101" on that core so we could read it later. Since the cores on the rope are already polarized as ones, we bypass the core to store the first zero. We would connect a sensing wire to

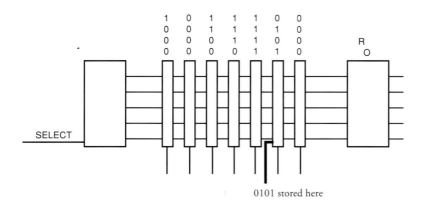

0101 stored here

Schematic of an abstracted core rope memory, storing only four bits per core.

our chosen core for the first "1," bypass the core with the second sensing wire (for a zero), and connect the third (for a 1). That way, when we selected that particular core, "0101" would be read out in parallel. The erasable cores had to be read *in serial* (one bit at a time), so they were much slower than the all-at-once readout of core rope. Since the Apollo core rope could accept up to 64 sensing wires, four computer words could be stored on each core—as it turned out, 2,000 bits in a cubic inch.

Despite the advantages in density and speed, the core ropes gave NASA and the software developers nightmares. One problem was the lead time needed in the Apollo program. NASA wanted the software delivered to the spacecraft and training simulators months before flight, so the astronauts could get used to it in their final rehearsals. But the core ropes had to be more or less handmade; Raytheon, the manufacturer, needed the ropes' software months in advance of NASA's required date. Therefore, to follow an optimum schedule, the software had to be ready to go up to a year before flight. In an atmosphere of constantly changing requirements and primitive software-development processes, this was hardly possible. Worse, since the core ropes had to be literally hard-wired and then sealed, any error of any kind got permanently wired in.

The software that *did* get to Raytheon was subject to two types of errors in the manufactured ropes: programming errors and manufacturing errors. You can imagine trying to follow a map of hundreds of thousands of ones and zeroes, and wire the correct "ones" to the correct cores. Once in a while, there was a "dropped stitch." Raytheon had to compare the core ropes against the software bit for bit, to make sure the missing or wrong bits did not get on the spacecraft.

The programming errors were more difficult to find and repair. Many times they were not found until after the software was in use. The Apollo programming team was feeling the pressure of trying to beat the Russians to the moon. When they got behind schedule, they took what they thought were safe shortcuts. Naturally, bugs got past the shortened development and testing.

Some defects were the kind made by inexperienced programmers working in a new domain. For instance, a particular type of mid-course engine firing for navigation required the astronauts to take some star sightings. To do this, they had to be out of their acceleration chairs floating at the eyepiece of the star sight. Once the computer had the data, the software did the calculations and fired the engines automatically. Unfortunately, the programmer did not build in a delay, so it was impossible for the navigator to get back into his seat before the engines fired. It is disorienting—and dangerous—to go from zero-g to accelerated motion without a chance to sit down. Even so, astronaut Gus Grissom was willing to put up with it. He said, "Well, we'll just lie down on the floor!" (There would be a "floor"—and a "down"—only as long as the acceleration lasted.)

Another error showed up in astronaut training simulations. On January 23, 1967, during a simulation of the upcoming first manned Apollo mission, the spacecraft computer and the ground computers were 138 seconds apart in their calculation of the time for a re-entry engine firing. This is the difference between landing in the assigned site and landing in, say, Zaire! It was mere weeks before the flight. Investigation showed that the ground computer's solution was correct. So, with the core rope installed and no time to make a new one, controllers told the crew to ignore the Apollo computer calculations and enter numbers sent up from the ground.

This workaround never had a chance to be used. On the 27th, a fire swept through the Apollo Command Module on the pad, killing Gus Grissom, Ed White, and Roger Chaffee, all experienced astronauts. During the nearly two-year delay in manned flights that followed the tragedy, the MIT programmers improved their software and its development process so that such errors would not haunt an actual mission.

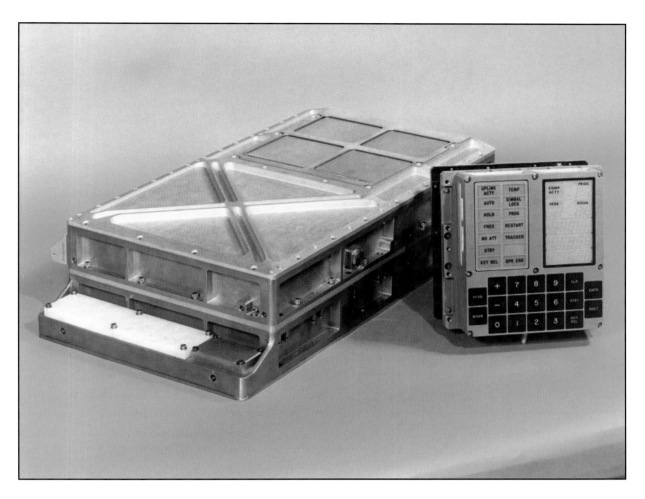

The Apollo Guidance Computer and a DSKY.

Flying the Apollo Computer

What exactly did the Apollo computer do? It provided some "services" (for example, acting as a clock to coordinate other spacecraft systems), but most importantly, it was the heart of the guidance and navigation system. Once the big boosters were gone, the spacecraft would be (for the most part) floating in free fall; guidance and navigation would consist of knowing where you were, and—very occasionally—firing the maneuvering engines to make small corrections in your orbit.

For instance, the S-IVB rocket stage put an Apollo from Earth orbit into the elongated orbit needed to intercept the moon. As the engine of the S-IVB ran, the navigation system measured its acceleration and calculated velocity changes. Then, just like a ship's navigator on Earth, the Apollo navigator took a series of star sightings. These gave position. The computer would then compare the spacecraft's position and velocity with the precalculated optimum position and velocity; a small maneuvering-rocket burn could bring them closer together, if needed. All this could be done automatically. The computer would change the attitude of the spacecraft and fire the engines for just the right amount of time.

It could also help when astronauts maneuvered the spacecraft manually. The computer in the Lunar Landing Module was the link between the pilot's

Apollo-Saturn V in flight.

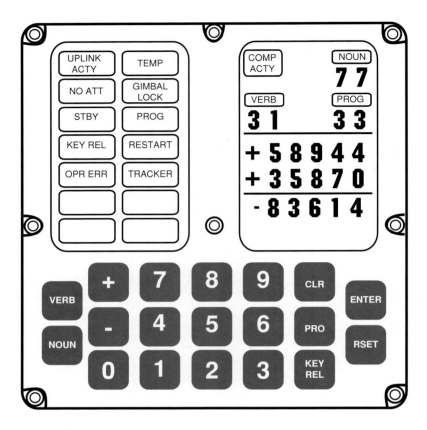

Detail of DSKY.

inputs and the module's attitude and control systems. For the moon landing, the astronaut looked out the window and controlled descent with throttle and stick, the same sort of controls as in an airplane.

Still Not QWERTY

The interface to the computer itself was not much like the keyboard-mouse-and-video-monitor we use today. If you look at the illustration of the *display and keyboard unit* (*DSKY*, or "disky") of the Apollo computer, you can see there is not much resemblance to the "QWERTY" keyboard used on typewriters and PCs.

The DSKY had three major areas. At the upper left, a set of warning and status lights helped "cue" the crew to things they needed to do, or would tell them something was happening. For instance, the "PROG" light meant the computer was waiting

for more data to complete an input to the current program. "OPR ERR" meant the astronaut had made a keystroke error that the computer could detect. "UPLINK ACTY" went on when data came up from the ground computers.

To the right of the lights, a display area showed computer operation. The "COMP ACTY" light blinked when a program ran. The two digits under "PROG" showed the number of the program that was running. A typical flight load had dozens of programs. The crew had to remember which was which. (For instance, programs numbered in the 30s were for rendezvous and maneuvering. Those in the 50s were descent programs.)

The other two sets of two-digit numbers, the "VERB" and "NOUN," were how the crew communicated to the software. There was no alphanumeric keyboard of the sort we use on PCs today. Instead, there were up to 100 verbs and nouns to use to make up commands such as "load program," "display time," and so on.

Let's say the crew wanted to load a rendezvous program and use it. They had a "cue card" for each program that told them which keys to press to get things rolling. Using the keypad at the bottom of the DSKY, they entered "VERB," "3," "7," "ENTR" to tell the computer that they wanted a program change. Then they entered "3," "1," "ENTR" to load P31, the rendezvous program. While P31 ran, the astronauts could ask for data—like the velocity change necessary for the next maneuver—by pressing "VERB 06 NOUN 84."

This was the same data shown on the Incremental Velocity Indicator in the Gemini spacecraft (see Chapter 5), except that Apollo put these numbers on three five-digit readouts, instead of on three three-digit displays. Also, the Apollo DSKY could show plus and minus before the numbers, instead of using "9" as the Gemini Manual Data Insertion Unit had done. Also gone were the arrows of the Gemini Incremental Velocity Indicator. The five-digit displays of Apollo could also show angles, as well as decimal numbers, octal numbers, and time. The astronauts could only tell what type of data was on the display by paying attention to context and referring to the checklist.

You would think that with 70 or so numbered programs, a couple of hundred verbs and nouns, and a display that could be in either of two different number systems (and a couple of different conceptual systems), an astronaut would be hopelessly overburdened by the computer. It did not turn out that way. With all the hours of simulations and rehearsals, the crew played the keyboard like a piano. One astronaut, Gene Cernan, said that when he pressed the wrong key it immediately "felt" wrong.

On the keypad were a couple of keys that modern computer users would have found strange. You can figure out that "RSET" is a way to get a "warm boot," and that "CLR" erases the current command, but what are "KEY REL" and "PRO?" "KEY REL," or *key release,* was how the computer got control of the keyboard. Unlike PCs, which have small programs that monitor the keyboard while it waits for keystrokes, the Apollo software had to "ask" explicitly for keyboard control, or pass it back to the crew by using the "KEY REL" light and button.

The "PRO," or "proceed," key was a good idea that turned into a pain in the neck. The astronauts had tried consistently to influence the spacecraft designers to give them more control whenever possible. The MIT software team tried to comply by

having the programs stop at certain critical points—such as just before an engine firing or major attitude change—and have the crew "approve" the next step. This worked by having the computer blink its lights to get the crew's attention. The astronauts would then have to look at the display, note which program was in progress, remember whether the output was decimal, octal, or something else, and then decide if it was OK to go ahead. If so, the crew pressed the "PRO" key.

This was not too much of a burden until the first real uses of the rendezvous programs. The lettering on the "PRO" keys was being worn off by the constant pounding of spacesuited fingers. The maneuvering and position updating were more frequent than expected. Also, since the crew was very busy *flying,* the tendency was to press "proceed" without thinking much about intermediate data. The solution was to have an alternative to most of the programs that reduced the number of astronaut keystroking. MIT called this option "minkey," for "minimum keystrokes." Once a program booted and was running, you could select the "minkey"

Part of a checklist used by Apollo astronauts working with the computer.

```
        P3I NCI TARGETING

        V37E 31E
        (If no REFSMFLG, to 3)
F 50 25    00017 MINKEY OPTION
    (Accept) PRO
    (Reject) ENTR

        SC CONT - CMC
        CMC MODE - AUTO
        (If req'd Mnvr <10 deg, to 3)
F 50 18    REQUEST MNVR TO R,P,Y ANGLES    (.01deg)

    (Accept) PRO
    (Reject) ENTR, to 3

  06 18    MNVR IN PROGRESS            (.01deg)
           When Mnvr Complete:  MINKEY, to 3
                                Non-MINKEY, to 2
F 06 95    TIG(NC1)              (hrs,min,.01sec)
           Load data if needed
           PRO
F 06 57    HALF REVS (NC1/NC2), (+0000N,.1nm,.1nm)
               ΔH(NCC),ΔH(NSR)

           Load data
           PRO
F 06 37    TIG(TPI)              (hrs,min,.01sec)
           Load data
           PRO
```

option by pressing (what else?) "PRO" at a specified point.

When Things Go Wrong

The software for Apollo matured rapidly after the 1967 fire that delayed the whole program. There were few in-flight computer problems. Perhaps the most serious of these happened during the very first landing mission, Apollo XI.

Slicing Time and Pre-empting Priorities

Usually a real-time control system (like that of Apollo and its successors) executes its software in one of two ways: by running several programs in pieces of fixed duration—or in the order of their priority. In general, at any given time during an Apollo mission, several programs would need to be run more or less at once. During descent to the moon, for instance, one program ran the engines, another measured acceleration and calculated position, and so on.

The programs were written so that they did enough calculations to accomplish their purpose, and then would "rest" for a few thousandths of a second while others ran. One way to achieve this

taking-turns arrangement was to give each program a fixed slice of time in which to run. Each would do as much as it could in its allotted time, and then be suspended (saving its data), while the next program in line ran.

The trick was to make sure the total length of all programs in the round-robin was short enough that something disastrous would not happen while a program was suspended. (Many simulators make certain that all their programs are executed at least once in every 40 milliseconds, or 25 times a second.) The advantage of using such *time-slicing* was that every program got some time to run. The disadvantage was that during certain periods of high load, a program's time slice might not be enough for smooth and safe operation.

The other method of running multiple programs in a real-time system is to execute them in order of priority. Let's say there are three programs on the current list, and they are of priorities 1, 2, and 3, where "1" is the most important. At the start, the executive program controlling the system runs the "Priority 1" code. When it is done, it runs the "Priority 2" program, then "Priority 3." If there is still time in the *execution frame* (clock period) defined for the cycle, it goes back to the Priority 1 program. The Apollo software designers used this method to run the Apollo flight programs.

Apollo XI's Lunar Landing Module left orbit and descended toward the moon early on the morning of July 20, 1969. Neil Armstrong commanded, and Buzz Aldrin was the pilot. The crew made several maneuvers, started and stopped flight programs,

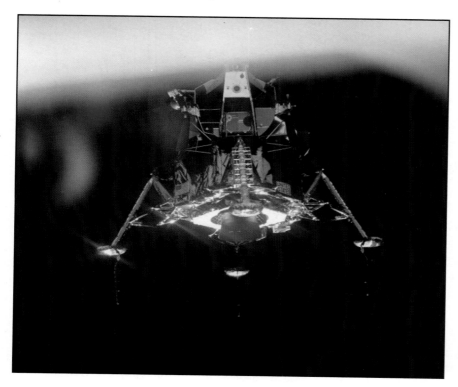

Apollo XI's Lunar Module, separated and maneuvering for descent to the moon.

and followed the instructions on extensive checklists. One item was deploying the rendezvous radar, which would be needed to find the Command Module if the landing had to be aborted. The signals from the rendezvous radar were analog—continuous voltages. The computer needed data in digital form, so the circuits between the radar and the computer included an analog-to-digital converter. This device would help cause the problem.

As Armstrong looked out the window for the planned landing site, Aldrin watched the instrumentation. Suddenly, a restart warning light went on. A few seconds later, it came on again, and then yet another time, all in less than 40 seconds. In the meantime, Armstrong saw the landing zone, but it was filled with boulders!

The photos from the earlier unmanned lunar orbiter could only show so much. On the basis of what they had seen, the mission planners had picked a site with numerous rocks—big enough to tip the lander over if a landing pad came down on top of one. That would doom the astronauts to a permanent stay on the moon. Armstrong looked for a better site, but it meant using precious fuel—and the computer problem did not seem to be getting any better.

What was happening? The execution frame for Apollo was 20 milliseconds. Normally the descent programs would execute within those 20 milliseconds, in order of their priority. As the software needed to add functions, it would put more small programs on a waiting list. At the end of each frame, the executive program would look at the list of waiting jobs, and check to see if one had a higher priority than the ones still being executed. If so, it would be added. As programs terminated, the 20-millisecond *priority interrupt cycle* would take them out of the executing group.

One design decision had the potential to mess up this system. The Apollo computer did not have as many *registers* in its arithmetic logic unit (areas in which to store temporary data) as modern computers would eventually have. It had to put information from its counters in its erasable memory—

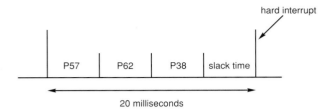

The 20-millisecond frame used for scheduling Apollo software.

which took time, and also required the attention of the central processor. The software designers had decided that requests for *counter increments* (updates from the counters) had the highest priority. When one of these came to the attention of the executive program, it would suspend the current job, handle the increment, and then put the suspended program back in action.

And Counting . . .

Many times a computer is asked simply to keep count: the number of times through a loop, the number of measurements made, the number of seconds left in a time interval, and many, many more instances. In Apollo, the current value of any particular count was kept in erasable memory. That way the software could be programmed to look in a specific place for a particular count, or would know where to add a new increment.

But what if there were so many increment requests that the suspended jobs *could not run* until it was time to begin them all over again at the start of a new 20-millisecond frame? If they were stopped from running for too long, they would stop servicing the control system—and the astronauts would effectively be riding in a falling rock. So the system did a restart to keep the software running long enough to keep the spacecraft flying. That was the restart warning light Aldrin kept seeing.

Fortunately (as described in Chapter 4), ground controllers had seen this situation in simulations, and had recommended the crew press on to landing. Armstrong found a parking space just as fuel nearly ran out.

Computers in Space

One of the defining moments of the Space Age: Buzz Aldrin on the moon, near the lander and the U.S. flag, July 20, 1969.

integrated into the control system that the only way to maintain control was with another computer.

The solution was the MARCO (for *MA*n *R*ated *CO*mputer) 4418. Built by TRW, it weighed "only" 32 pounds and required 90 watts of power. It came in handy for the astronauts as a source of navigation data when the lander maneuvered on the far side of the moon, out of radio contact with the ground computers. It could fly the ascent module of the lunar lander (leaving the landing legs and descent engine behind) to a rendezvous with the Command Module—but nothing else. While it was never needed for its original mission (to take over in the event of an aborted landing), the Apollo XI LM crew used it to help them get back to the Command Module, just in case the primary system had more problems.

Post-flight analysis showed the source of the trouble. A complicated—and unforeseen—routing of the signals from the rendezvous radar had put the analog-to-digital converter in a "race" condition. In response, it sent excessive requests to the executive program, resulting in too many interrupts. At one point, the LM's computer had devoted *15 percent* of its computational resources to handling the rendezvous radar. The fix was to change a switch position—a change duly added to the checklist for the next flight.

What would have happened had controllers told the crew to abort? In the Command Module, failure of the guidance computer meant the ground computers would become the primary source of navigation information. The astronauts would have had to fire engines and point the spacecraft manually, just as in the Mercury and Gemini days. The margin for error in those spacecraft was not great, but it was within human limits. A computer failure in the lander was a different story. There would have been no time for an update from the ground—and the computer was so completely

Twilight of Apollo

The lunar landings ended in 1972, only three years after men first set foot on the moon. No one (at least, no one from Earth) has been back since. The Apollo hardware, built at such cost and human effort, has been sent to museums.

NASA flew Apollo spacecraft four more times in the early 1970s. Three missions ferried astronauts to the Skylab space station for stays of one, two and three months. The fourth was the "Apollo-Soyuz Test Project." A Soviet Soyuz spacecraft orbited as a passive target, much like the Agena had done for the Gemini missions, and an Apollo Command Module hooked up with it for a couple of days in 1975. The mission provided practice (supposedly) for potential cooperative and space-rescue missions. Nearly 20 years would elapse, however, before an American Shuttle would be slated to fly to rendezvous with the Russian *Mir* space station. Still, in the midst of the Cold War, Apollo had ended with a spectacular gesture of hope.

As the Apollo program phased out, the simple ballistic shapes of the first three U.S. manned programs gave way to the blocky winged shape of the Shuttle. No Mercury, Gemini, or Apollo spacecraft had ever flown twice. Neither had any significant hardware, whether rocket or computer.

The demands on the new spacecraft would be different. The Shuttles were designed to fly dozens of missions over their operating lives (which now will probably stretch to 2030—nearly 50 years of use). Their solid-propellant boosters could be recovered when empty, and refueled. The liquid-propellant engines would also be reusable. Only the relatively cheap external fuel tank would be discarded after each mission. The Orbiter portion of the Space Transportation System—the Shuttle itself—would be designed to fly a lot like an airplane—and it would only be able do so under computer control. On-board computers had become not only integral, but essential to space flight.

An astronaut explores a large crater on the moon with a lunar rover (background), in the later stages of the Apollo program. There have been no manned flights to the moon since December 1972.

Computer Technology for the Shuttle

7

During the powered ascent to orbit, the pilots of the Shuttle do much as their Mercury, Gemini and Apollo predecessors did: they monitor the flight and watch for anything that might go wrong. The Shuttle astronaut-pilots have a big advantage: three display monitors in the front cockpit give a wealth of information about spacecraft systems, and many command options.

The software on the Shuttle can handle complex procedures. These encompass *ascent processing* (calculations during the climb to orbit), *on-orbit operations* (such as deploying satellites and interplanetary spacecraft like *Galileo*), the difficult descent to Earth, and landing on a normal runway instead of a point in the ocean. But this capability is no luxury.

One major difference between the Shuttle and some earlier spacecraft is its total dependence on the flight computers. There is no mechanical or electronic backup. If the computers have a catastrophic failure, then the astronauts die. Why is NASA is willing to take such a chance on a single system? To find out, we must follow a surplus lunar module computer in a journey to Edwards Air Force Base, location of NASA's Dryden Flight Research Center.

Computer technology helps a Shuttle launch go smoothly.

The test orbiter *Enterprise* is shown here atop its Boeing 747 carrier aircraft, with NASA's modified F-8 Crusader fly-by-wire aircraft in the foreground. This F-8 helped prove key flight-control technologies used in the Shuttle.

Computers in Flight Control

When people first fly in a light plane where they can watch the pilot, they are often surprised at how little effort goes into controlling the plane while cruising. Once a typical airplane with a mechanical control system is leveled off and the power adjusted, the pilot sets the "trim." This setting compensates for the weight distribution of passengers and cargo. A relatively narrow tab, attached to the trailing edge of the *horizontal stabilizer* at the tail, is moved by the trim wheel, deflecting into the slipstream. As long as power is not reset and no one changes seats, the plane stays (for the most part) on the straight and level.

It was not always like this. In fact, almost all early airplanes were unstable. Anyone who tries to make that "perfect" paper plane knows how difficult it is to have one fly on its own, nice and steady. This is the problem the Wright brothers and their competitors had as they tried to design the first powered airplanes.

The Wrights solved the control problem without actually balancing their airplane. At first, their gliders had wings and not much else. The pilot stabilized the glider by shifting his weight. That was difficult—even on non-windy days—and it was not a long-term solution.

As the Wrights made better gliders, they added two *vertical stabilizers* in the back of their biplane, and two horizontal stabilizers in the front. They also connected wires to the trailing edges of the wings so the pilot could "warp" the wings in flight, and make them dip one way or the other. These additions to their contraption helped the pilot maintain control. If the plane nosed over forward, the horizontal stabilizers had to point up to recover. If it leaned to one side, warping the wing righted it.

This concept of *active control* was not really too difficult for the Wrights to grasp: they also made bicycles, one of the most unstable vehicles ever built. Even so, a child can ride a bike; and pioneer pilots soon got used to the foibles of their airplanes. Or they would die trying.

Throughout World War I, most planes were unstable. The darting aerobatic biplane and triplane fighters of that war actually benefitted from their instability. They were automatically more maneuverable in the direction in which they were unstable, so they could move rapidly and escape that way. Also, the tiny fuel tanks of an early airplane meant it could fly for only a few hours. The pilots

Control Surfaces

All movement of an airplane or spacecraft is identified by referring to the position of the *longitudinal axis*, the one running lengthwise through the center of the vehicle's body. The control surfaces of an airplane move it in three axes. On a spacecraft in a vacuum, it is small control jets that move it in the same three axes.

In the case of an airplane, the elevator moves the craft in the *pitch axis* (nose up and nose down). The rudder moves the airplane in the *yaw axis*. Finally, the ailerons move the airplane in the *roll axis*.

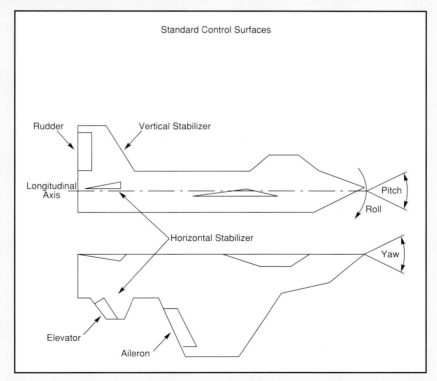

The three axes of flight control (pitch, yaw, and roll), and standard aircraft control surfaces.

would not get overly tired wrestling with the airplane in that length of time.

After the war, airplane builders began to concentrate on range and on passenger-carrying capacity. On long flights, with passengers wanting comfort, stability was desirable. As engineers became more consistent at designing stable *airframes*, the unstable ones more or less disappeared. Maneuverability in World War II aircraft came from good control surfaces and sheer power, rather than lots of lift and instability.

As the end of the Second World War approached, several new high-performance aircraft and unique configurations came out of the design bureaus of the Axis and Allied Powers. There were jets, flying wings, rocket-powered aircraft, and rockets themselves. Many of these had severe control problems that had to be solved. As we saw in Chapter 2, Helmut Hoelzer's guidance computer could stabilize the V-2 rocket. In jets, additional hydraulics (like the power steering in your car) helped pilots maintain control.

One goal of postwar aeronautical engineers was to break the "sound barrier." Conventional piston-driven planes with propellers had experienced all kinds of trouble as they approached *Mach One*, the speed of sound. Early jet pilots also discovered some rather nasty control effects as they neared the threshold of that speed—which really did act like a "barrier." The reason for this is that as Mach One comes closer, the airplane's *center of lift* actually shifts toward the rear. Prior developments in flight control had not prepared aircraft to handle these new conditions.

The Wright biplane leaps skyward in powered flight, December 17, 1903.

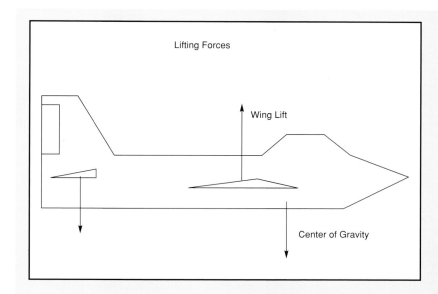

Lifting Forces

Wing Lift

Center of Gravity

The lifting force of an aircraft's wing must counteract the pull of gravity, requiring a delicate balance. In transonic flight, the aircraft's center of lift changes.

The Moving Lift Factor

In the first figure, the force of lift from the wings balances the pull of gravity (shown as a downward force), along with the downward force of the horizontal stabilizer. In the second figure, *transonic* flight—the speed at which the aircraft exceeds the speed of sound—the force of lift is moving to the rear of the wing, putting the finely-balanced forces out of whack.

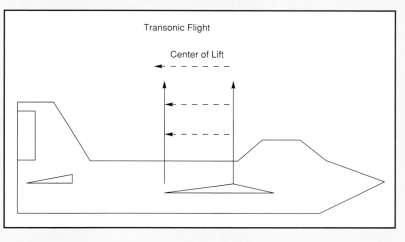

Transonic Flight

Center of Lift

When the designers of the 1920s had searched for stability, at least they had made the horizontal stabilizer more than "just" a control surface. The Wrights had derived some stabilizing effect from their little wings in front of the big wings, but mostly these helped point the flying machine. They were, in fact, all of one piece and "all moving"; the entire stabilizer moved when the pilot pushed or pulled the control lever. Later stabilizers took on the common form they have today: a solid piece with a moving section at the trailing edge, mounted on the tail. The air flowing back from the prop would "push" down on the solid part of the stabilizer, counterbalancing the heavy engine in the nose. If the pilot wanted to move the nose

up or down, control cables attached to the stick in the cockpit adjusted the angle of the moving part of the stabilizer, the elevator.

When flying wings, delta wings, and jets (which had no prop wash) came into use, control surfaces had to be different. When a finely-tuned design approached the speed of sound and the center of lift shifted, the control responses had to shift with it so the pilot could keep control.

Along with high-speed airplanes came more problems in making them maneuverable. Trying to "power through" turns meant greater stress on wings and other surfaces. Making these stronger meant they would be heavier, which worked against the goal of lifting more payload.

107

General Dynamics F-111 prototype in flight, with wings at about half-sweep.

Test Aircraft Are Both "Old" and "New"

Most people think of the flashing X-series aircraft when they think of test airplanes and test pilots, such as Chuck Yeager and the X-1, the X-15 rocketing to the edge of space, or the X-29 and its funny-looking forward-swept wings. Most test airplanes, however, are modified or obsolete craft that can still be used to demonstrate or try out some new concept. Such were the retired B-47 bomber and early F-8 fighter used in fly-by-wire research. The term *testbed* is sometimes used in reference to this practice, as in "fly-by-wire testbed," or "B-47 testbed."

In the 1950s and 1960s, the Air Force began to do serious research in *active flight control*. This meant a return to unstable airframes. Since the planes would be unstable anyway, the shift in lift at supersonic speed would be solved just by solving the control problem in general. Plus, the maneuverability of the planes would be automatically increased in the unstable axes.

By the mid-1960s, there were already some partial solutions flying. The F-111 needed to solve complex control problems in the development of its "swing wings." (Imagine the changes in center of gravity and lift as the wings pivot out for greater lift, or in for greater speed.) The designers had put in a "stability augmentation" system to give the pilot automatic help in keeping control. These stability-augmentation systems already had small analog computing circuits; perhaps such a setup could work for active flight control. As a start, the Air Force decided to modify an old B-47 bomber to use slightly more sophisticated computers in the pitch axis of its flight control system.

In the typical B-47, cables connected the control stick to the elevators. Moving the control stick moved the cables, and they deflected the elevator. The Air Force engineers left the cables in place as a backup, but they also ran electrical wires from the stick to an analog computer, and then from the computer to electric motors at the hydraulic system that moved the elevator.

When the pilot moved the stick, the pressure he put on it was transformed into a voltage in proportion, and the computer took that as input. The computer also got information from sensors that could tell it the attitude of the airplane. The computer took the two sets of data (the desires of the pilot and the current state of the airplane), compared them, and calculated commands to the electrical servos. This is a lot like the way the Apollo computer worked when it maneuvered the spacecraft. The test system in the B-47 worked so well that the Air Force moved forward, trying a control system for all three axes of flight (pitch, yaw, and roll) in a modified F-4 fighter.

Meanwhile—completely independently of the Air Force—NASA engineers at the Dryden Flight Research Center also had the idea of trying to build one of these active-control, or "fly-by-wire," control systems. They obtained a surplus Apollo guidance computer to use as the heart of the system. An analog system like the one the Air Force used in the F-4 would be mounted side-by-side as a backup. NASA wanted to use the digital computer because they recognized that if something was not working well, altering software was easier than altering the entire hardware of the control system. Besides, they had a good software contractor at MIT who could help them with the new control programs.

Ironically, both the Air Force's modified F-4 and NASA's F-8 flew fly-by-wire independently, in the same month: April 1972. From that date, the technology of flight control changed forever. By the end of the 1970s, the Air Force fielded the F-16 lightweight fighter, which used analog computers in much the same way as the F-4 testbed. In the early 1980s, the later-model F-16s, the F-117 stealth fighter, the F-18, and the Airbus 320 commercial transport all used digital fly-by-wire control systems. By the early 1990s, all significant new aircraft were fly-by-wire: the F-22 advanced tactical fighter, the B-2 stealth bomber, the C-17 military transport, and the Boeing 777 commercial airliner. All of these latest aircraft use digital flight control *exclusively*. There is no mechanical system backing up the computers. This technology owes a great deal to the Shuttle program, which had to solve the "reliability problem" first.

The B-47 modified for use by the Air Force as a fly-by-wire testbed.

Computers Fly the First Space Station

There are two ways to make any system using computers more reliable. One is carrying a back-up, the other is *redundancy* (mounting one or more duplicates). Through most of the space program, NASA has chosen to use backups; they can be made simpler than the primary computers. They do not have to do quite so much. We saw in the chapter on the Gemini spacecraft that the ground computers were the navigation backup, and the pilots themselves controlled maneuvers. In the Apollo lander, there was a small computer that could perform one basic function: abort a landing and get back to lunar orbit. Since there was a single computer on NASA's experimental F-8, there had to be an analog backup.

However, there are times when a mission is so long—or abandoning the main objective would be so expensive—that a fully redundant system is needed. Redundant systems fully replace the functions of the primary. We will see that the Jet Propulsion Laboratory used redundant computers in all its long-duration space missions. The first manned spacecraft to do so was Skylab.

Skylab was the first space station. NASA took the third stage of a Saturn V and modified it to serve as an orbiting workshop. It had a big solar telescope and dozens of experiment stations, plus living areas for the crew of three. The plan was to send up teams of scientist-astronauts for ever-increasing stays in the station. It was to stay operating for years, but would not always have a crew on board.

Since Skylab would be up for so long, using attitude-control jets to point it one way or the other was not possible. The jets would need too much fuel. As an alternative, a NASA engineer designed a gyroscopic system that could change Skylab's attitude.

The gyro system required a powerful computer. If Skylab were turning end-over-end, it would be difficult for the Apollo command modules to dock with it. Therefore it had to keep Skylab stable even when no astronauts were aboard. The computer chosen was a powerful IBM model that had the same architecture as the famed System 360 mainframe, but (of course) was a lot smaller. It used a 16K core memory and had a tape drive with back-up programs, much like the Gemini computer.

Since the computer had to be constantly working, NASA decided to use two completely redundant machines loaded with identical software. They knew an IBM service call to low Earth orbit would be pretty expensive!

Whenever there is a redundant computer system, the most important thing to know is *which* of the com-

The Apollo Guidance Computer mounted in the left gun bay of the NASA F-8. Note the DSKY used to start the software before the airplane could fly.

puters is operating correctly. You do not want the failed computer to be in charge when a perfectly good computer is running right next to it. *Redundancy management* is the key. IBM's suggestion for Skylab was the construction of a voting device that was itself redundant.

IBM designed a set of circuits using "triple modular redundancy." This meant building each critical interface in sets of three. Signals passed through all three modules, and also the voting modules placed at intervals. Any data that got two of three (or three of three) votes passed on to the next stage. In flight, each computer ran a self-testing program at the beginning of each real-time processing cycle.

It was like setting the two computers down opposite each other, with a math exam in front of them and an on/off switch between them. Each cycle they would do a problem on the math exam, and look at each other's answer. If the answers didn't match up, the computer that detected the error would turn off its partner.

Of course, there was a small danger in this arrangement: if Computer A considered Computer B loony, then Computer B would probably not be thinking so well of Computer A, either. The wrong computer might somehow get "killed." As a safeguard, a timer circuit in the triple modules monitored the "survivor" of an error, and switched it off if it did not reset the timer within a certain interval. A failed computer was hardly likely to be doing such housekeeping.

All this preparation was for naught: the Skylab computers never failed. In fact, they were "heroes" on more than one occasion.

A military application of active flight control in the 1990s: The Lockheed/Boeing F-22 Advanced Tactical Fighter.

Computers to the Rescue

When the Saturn V carrying Skylab roared into space on May 14, 1973, some external shielding tore away as the rocket went through the atmosphere. One of Skylab's big solar panels ripped away; the other was trapped by a snapped cable. The first telemetry from the space station was not good: low on power, gyros drifting, internal temperature rising.

The first crew was supposed to fly up the next day, but now they would have to wait until repairs could be planned. In the meantime, the exposed parts of the station had to be shielded from the sun, or the fragile electronics inside would be baked to uselessness. The only way to protect the space station—and to keep power coming from its telescope-mount solar array—was to adjust its attitude constantly.

With no human crew on board, the TC-1 computer took over, helped by a continuous stream of instructions from ground controllers. It was two weeks before the human beings showed up—to find no real damage to the spacecraft systems, other than what had happened during ascent.

Years later, the computers got to be boss of the space station again. The original Skylab mission

lasted 272 days, with about 180 days of astronaut occupancy. When the last crew left, they turned everything off, including the computers. By then a sunshade protected the main part of Skylab; the station went into a long sleep, without having to worry about drifting. The money had run out. NASA intended to boost Skylab to a higher orbit once the Shuttle began flying. Maybe then they could use it as the core of an enlarged space station, adding on until it looked much like the Russian *Mir* does now.

Unfortunately, the middle 1970s was a period of greatly increased solar activity. As the sun blasted Earth with different kinds of radiation, the atmosphere thickened. Even at hundreds of miles above Earth, the drag of the atmosphere slowed Skylab. Its orbit began to decay. NASA realized there was a danger that tons of melted metal could survive reentry, endangering people and property on the ground. They prepared two new computer programs. One pointed the narrowest part of the space station in the direction of flight, to reduce drag and help it stay up longer. The second program could exercise some control over re-entry, and might prevent Skylab from crashing through some unfortunate suburban family's roof.

Before the programs could be loaded on Skylab, engineers needed to find out if the computers still worked. In March 11, 1978, they briefly fired up the processors. The computers came on line, exactly where they had left off four years and 30 days earlier! (This is the best thing about core memory.) Four IBM programmers coded the two new control codes, and they sent them up to Skylab. The on-board computers retransmitted them bit-by-bit to the ground, where they were manually checked. Once IBM and NASA were satisfied, the new programs kicked in.

For 393 days the computers flew Skylab, trying to save it for the day the Shuttle could come to the rescue. (Unfortunately, the Shuttle program at that time had suffered all sorts of delays.) Finally, the re-entry program had to be used as Skylab burned up. A big piece of a tank and other scattered fragments landed in the Australian outback. The largest remnant, as big as a small car, rests in the Alabama Space and Rocket Center in Huntsville—not far from NASA's Marshall Space Flight Center, where the Skylab project originated a quarter century ago. The computers, inanimate heroes, had gone down with the ship.

Four Computers Marching in Step

Skylab proved that redundancy was a workable option for space flight. The Shuttle made it a necessity. The designers of the Shuttle felt that the most likely points of failure lay in the avionics system—and the computers would be the heart of that system.

It would be impossible to use only two computers for flight control; the task was more complex, and the stakes were higher. If Skylab's redundancy-management software had picked the wrong computer during a failure, the worst that could have happened would have been a space station pointing the wrong way until the crew loaded backup software and rebooted the system. On the Shuttle, an undetected or mismanaged computer failure during powered flight or descent would kill the crew.

Adding a third computer would improve things somewhat, because then voting would be more meaningful. Two computers failing at the same time was unlikely. But if one did fail, it would be back to two computers again—a bad situation. Therefore a better option would be four, and five even better than four. By specifying five computers for the Shuttle, NASA achieved its desired level of redundancy:

fail operational/fail operational/fail safe

Skylab, the first space station, in orbit. Note the missing solar panel, and the metal foil "sunshade" installed by an astronaut repair crew.

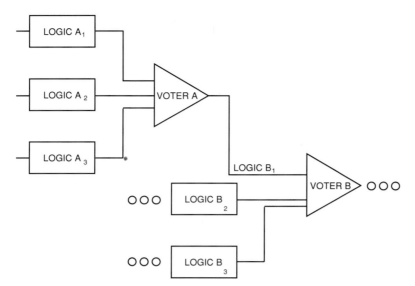

Schematic of a triple-modular-redundancy logic circuit.

This specification meant: "One computer failure and you are still fully operational, a second and you are still operational, but then it's time to come home."

Such deep redundancy had not flown before. Fortunately, NASA's F-8—the digital fly-by-wire testbed—had finished its single-computer flights, and the engineers at Dryden wanted to take the next step in active flight control. They knew a single-computer control system would never be acceptable even to the military, let alone the commercial airlines. For Phase II of the F-8 program, they wanted to try multiple computers.

Shuttle program engineers at Johnson Space Flight Center knew of the fly-by-wire program. They counted on its results to provide help in designing and programming the Shuttle control system. The two groups of engineers talked things over, and decided to go together on the request for bids for flight computers.

Who Is Right? Who Is Wrong?

NASA and IBM searched for a better redundancy-management scheme than the one used on Skylab. Things quickly got more complicated. NASA

decided to relax the safety margin to "fail operational/fail safe," which meant the fifth computer would not be needed. Instead of eliminating it, they decided to keep it as the heart of a "Backup Flight System." That way they would get the best of both worlds by having a less-capable unit (as in previous spacecraft designs) to back up the four processors in the redundant set. This "best of both worlds" came at a cost of further work on integrating the new system.

The basic way of detecting a failed computer was to observe it out of synchronization with its identical siblings. The standard rationale held that if one computer says "turn left" while three others are saying "turn right," the most likely reason is that the majority is correct, and the dissenting computer ought to be turned off.

They both wanted off-the-shelf, existing computers. No more would they accept single-purpose machines that were difficult to maintain (and just plain expensive). By then the series of IBM computers that had included Skylab's TC-1 had matured further. NASA chose the AP-101 computer from IBM for the both F-8 Phase II and the Shuttle. The same model flew on the F-111, B-52, and B-1, mostly for navigation and weapons delivery. In the F-8 and the Shuttle, it would provide active flight control as well.

At first, the engineers thought they could detect the bad guys by starting all the computers at the same point in the software, and pausing to examine a *checksum* (the sum of the contents of all the registers in the central processor is one type of checksum). Oscillators, however, are the heart of a computer's clock—and anyone who knows how they are manufactured will also know that here was a problem: it is highly unlikely to find four so alike that they could keep time—exactly in sync—

at the microsecond intervals needed. The four computers quickly got out of sync by tiny increments; there was no way to tell who was wrong or who was right.

Computer in a Closet

Even though the Shuttle has a redundant set of four computers, and a fifth on line as a backup, it was not enough for NASA. During the first few flights of the spacecraft, there was a sixth computer in a storage locker. In case of an in-flight failure, the astronauts could remove the broken machine and replace it with the spare. The computers are all in the mid-deck avionics bays, easily reachable by the crew; they could disconnect a few cables and pull out the offender. When the Shuttle resumed flying after the *Challenger* disaster—with new-technology computers—the sixth computer once again flew on missions. Even though there is no excessive danger in flying with one computer out of service, the "computer in a closet" actually was used on STS-30, when the Number Four computer failed.

NASA assembled a team to study the issue, drawing personnel from Johnson Space Center, the Draper Laboratory (the new name of the Instrumentation Lab after MIT divested it), Rockwell International (the Shuttle prime contractor), and IBM. The team came up with a solution used to this day.

Each computer would run until it reached an input, an output, or an instruction to switch to a new program. When any of those things happened, the computer would send a three-bit message to all the other computers in the redundant set, along with a code for the operation. Then it would wait. Within four milliseconds, it would have to receive the same message from all the other computers. If another computer sent the wrong message—or no message—in those four milliseconds, the computer detecting that failure would vote the offending computer "bad," and refuse to listen to it in the future. The idea was that three computers would be voting the failed computer "bad" at the same time the bad one was insisting the other three were wrong. The majority would rule.

Unlike on Skylab, the computers could not turn each other off nor drop each other from the redundant set. Only an astronaut could do that. There was a good reason. If a software error in one machine caused it to turn another one off by mistake, the error would be present in *all* the machines of the redundant set—and they might *all* turn each other off.

At the commander's seat in the Shuttle cockpit (the left seat), a matrix of lights above the window shows signals in the event of a computer failure. If a computer is voted bad by the other three, the failure lights will point to it. The commander can

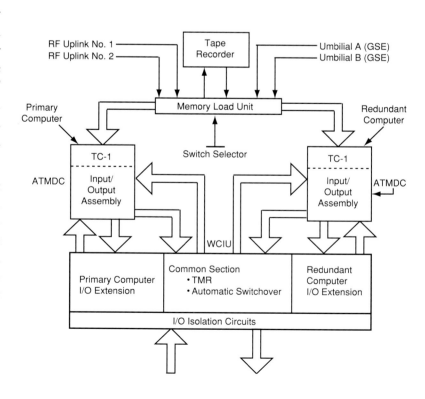

Schematic of the Skylab computer system.

reach up to a set of power switches and "kill" the offending computer.

This procedure takes time—eons of computer time, as measured in hundreds of thousands of instructions executed per second. Why doesn't the failed computer cause any damage to the mission? Remember the triple-redundant voter circuits in Skylab? They were right at the computer's output interfaces. In the Shuttle, the voter circuits are at the actuators that move the control surfaces. If conflicting computer signals cause force-fights between actuators, these are won by the majority.

Long before the Shuttle flights, this redundancy-management scheme had flown on the F-8 fly-by-wire testbed. It had successfully survived several in-flight processor failures. When a Boeing 747 carried the orbiter prototype *Enterprise* aloft to begin approach-and-landing tests, the computers could be tried out on a Shuttle in unpowered atmospheric flight. On the very first such test, a computer failed the instant the orbiter left the jumbo jet's back—and the new redundancy-management scheme had an immediate successful test.

The F-8 Phase II flight program provided direct help to the Shuttle landing tests at least once. During some landing approaches, the Shuttle suffered from what appeared to be "pilot-induced oscillations"—which happen when the pilot tries to recover from something the aircraft is doing, and overcompensates. For instance, if the Shuttle nosed over, the pilot would pull back on the stick. If he pulled into a near-stall condition—or some other position outside the computer's tolerance— it would force the nose *down*, causing the pilot to react again. The result would be a Shuttle doing a porpoise imitation in the sky.

To help solve the problem in an inexpensive way, NASA had the F-8's computers reprogrammed to act like the Shuttle in flight. (This is one great advantage of software and digital control systems: one airplane can be made to act just like another.)

The F-8 pilots replicated the Shuttle's problems in an aircraft that could be flown more frequently, under its own power. The results of their flights resulted in software changes that fixed the problems.

Different Phases, Different Configurations

While under power during ascent and in re-entry, the Shuttle's four primary computers run duplicate software. On orbit, one or two computers are turned off to save power. Then the redundant set might be only two machines, with a third doing system management and payload processing.

In the unlikely event that most of the computers in the redundant set fail during a critical mission phase, the pilots can press a button on their control sticks, and the backup flight system takes over. The software on the backup is for survival only— much like the backup on the Apollo lander. Using the backup software, the Shuttle can fly into a safe orbit and return to Earth—nothing more.

My Other Airplane Is a Spaceship

The Shuttle program has a Gulfstream corporate jet with a control system coded to act like the Shuttle. By using thrust reversers and the help of the software, a Shuttle pilot in the Gulfstream feels as though he or she is landing the real thing. The cockpit details include even a modified seat with a Shuttle hand controller.

The backup flight software is not part of the synchronization scheme. It runs on its own. In fact, it has a different operating-system philosophy. The primary computers use interrupt-driven scheduling, just like the Apollo software. On the backup, the programs are *time-sliced* (see Chapter 6).

Nevertheless, the backup has to be on the same page as the primary computers (that is, have the most up-to-date information) in case it is needed.

Therefore it has to be in step with the primaries when it comes on. In general, this was not too much of a problem—but the very first Shuttle mission had a four-day delay because of it.

When the computers came on during the countdown, the backup refused to sync with the others. This had not happened before. Puzzled, NASA and IBM engineers tried, over and over again, to bring the computers up in a simulator in Houston. It worked normally over 70 times. Finally, it failed—and they discovered a small error injected into the program over a year earlier. By this time, the countdown had been stopped and the flight rescheduled for a few days later.

When they tried to boot the computers on *Columbia* again, they came right up and synchronized. The engineers simply did not turn them off again, and fixed the bug after the flight.

When *Columbia* finally lifted off in April 1981, all the basic technology of active flight control and redundancy management came into play. The Shuttle program already had nearly eight years of software development behind it. Now the astronaut crews could fly the computers on the most complex spacecraft ever built. Eight and one half years later, as *Galileo* left the cargo bay of *Atlantis*, the Shuttle would retain that distinction.

Enterprise, shortly after separation from the Boeing 747 carrier, during Shuttle approach and landing tests.

Programming and Flying the Space Shuttle Computers

8

When STS-34 lifted off with *Galileo* on board, the Shuttle Data Processing System was running at full steam. The four processors in the Primary Avionics Software System and the lone Backup Flight System computer sent thousands of signals throughout *Atlantis*. A quarter-century of computing history rode into space.

Orbiter *Endeavour* in its giant sling, being lowered to mate with its solid rocket boosters and external fuel tank. NASA's newest manned vehicle, it has several technology improvements over its 10-year-older partners.

Hardware and More Hardware

NASA's AP-101 computers were based on an architecture IBM had originated in the early 1960s—almost in spite of itself. In the very early days of the electronic computer, IBM had disdained it as a product.

They'll Never Sell . . .

Thomas Watson, Sr., CEO of International Business Machines, reportedly thought the total worldwide market for mainframe computers would be five or so. Ironically, a later era would prove him close to right—but only because the market would explode into a new world of desktop workstations and PCs more powerful than mainframes of the 1960s.

How the word "mainframe" came to represent very large central processors is not clear. What *is* clear is that the average mainframe of the 1950s and 1960s filled over a thousand square feet of space, and had high requirements for power and maintenance. Only a few existed—at least relative to desktop mechanical calculators—so it was hard to imagine the market for them.

As Remington Rand's UNIVAC and other machines came to be applied to a variety of tasks, however, IBM finally entered the fray. During the 50s it developed several models of mainframe computer for scientific and business calculations. By 1960, the 7090 series entered service, and found use at NASA in simulations. The upgraded 7094 served as the Gemini control-center computer.

But the crowning glory of IBM's mainframe line in the 1960s was the System 360, a powerful computer built in many versions. These ranged from the low-end 360 Model 20 and 40 used in business applications, through the Model 67 (a pioneer "time-sharing" machine), up to the Model 90s used in giant government data-processing shops. IBM built the computer by first writing the "Principles of Operation," a manual that explained the instruction set and machine-level programming. Thus the hardware efficiently implemented the low-level software. It used instruction words of 16 and 32 bits, and could do floating-point arithmetic with as many as 64 bits.

IBM thought the 360 architecture could be used in a variety of packages. Beginning in 1966, they marketed it in the 4Pi series of computers for use in *embedded systems*. One of the 4Pi derivatives was the TC-1 that flew in Skylab. A later version, the AP-1, was the competitor put forth for use in the Shuttle. The actual Shuttle machines would be an upgrade of the AP-1, known as the AP-101. Many further upgrades followed; now the Shuttles fly with the AP-101S.

The original AP-101 used 16-bit addressing, so only 64K of memory could be directly accessed. This was fine; Rockwell International's analysis of Shuttle requirements had indicated that 64K would be plenty. NASA wanted the new computer memory to carry the entire flight load at liftoff, as that of Apollo had done.

On the basis of the preliminary analysis, NASA bought a series of these 64K machines. They flew on the F-8 fly-by-wire testbed, and on the *Enterprise* during the drop tests. Unfortunately, as the real flight software took shape, NASA found itself in the same situation it had survived in Gemini and Apollo: too much code, too few cores.

Computers as a Component

Embedded computers are processors and peripherals installed within a larger system. Most of the time we think of computers as standing alone—on a desktop or in a computer room. In an embedded system, the computer is an integral part of the entire operation. All on-board computers in spacecraft are examples of embedded systems.

There was a real (though it not strictly technical) problem this time: the Space Race was over. Money was not being thrown at the space program with the same enthusiasm as it had been a decade before. Buying computers for the Shuttle program was no small deal. Each orbiter needed five, each simulator needed five, each development computer needed a couple—and then there were spares, a total of up to 50 machines. Memory upgrades cost as well.

NASA stretched the budget and got 106K memories for the AP-101s, still too small to hold the entire flight load. It would be necessary to split up the load (as in the days of Gemini), and store the programs on good old magnetic-tape mass memories. Once tape memories became available, NASA decided to store every piece of needed software on them, from preflight checkout code to landing programs.

Since the AP-101 machine was originally an "off-the-shelf" computer—intended for a variety of applications—it required some modifications to match the needs of the Shuttle. The modified AP-101 design kept a small amount of programmable microcode available to allow changes, or the addition of a few special instructions. Also, wherever AP-101s were used, a customized input/output processor connected them to the overall avionics system. The resulting system still flies today.

The input/output processors on the Shuttle connected the general-purpose computers to many spacecraft systems. These included the

The tallest tree in the forest: *Endeavour* rides to the launch pad on a mobile transporter.

The upgraded AP-101 used as the Shuttle's primary avionics computer.

actuators that moved the control surfaces, the solid rocket boosters, and the liquid-propellant rocket engines.

Each processor required 24 data buses to do it. One of these—the *intercomputer communication bus*—carried the signals that synchronized the computers. There were other communication buses as well, and eight flight-critical buses for actually controlling the spacecraft. The computers and buses were arranged in "strings" which could be reconnected in various ways to provide the most safety and redundancy. Each subsystem had at least two bus connections. Each bus itself had a small microprocessor to control its own traffic.

The original AP-101 computers ran at just under one-half million instructions per second (.5 MIPS). (For comparison, most desktop workstations nowadays can process tens of MIPS, and soon 100-MIPS home computers will not be unusual.) Although .5 MIPS was an improvement on Gemini's seven thousand instructions per second, NASA wanted more processing speed. Early in the 1980s—after the pressure simply to get the Shuttle flying subsided—the agency discussed upgrades with IBM. They settled on a memory increase to 256K (the maximum that could be addressed without changing the architecture), the use of improved microcircuits, and semiconductor memories. NASA could not afford to change the architecture; such a move would have meant a massive change in the (incredibly expensive) custom software.

Power Steering for the Shuttle

An *actuator* is a device that uses either hydraulic pressure or electric motors to move control surfaces such as ailerons or rudders (see Chapter 7). These actuators are needed in large conventional aircraft because even Hercules would have trouble moving a one-ton control surface on a Boeing 747 in a 400-knot slipstream. In a fly-by-wire system like the Shuttle, they are essential, since there is no direct connection between pilot and control surface. The control signals come to the actuators on *data buses*, which are connectors between computers and devices throughout the spacecraft.

The newly-upgraded computers were more compact—together with their input/output processors, they were no bigger than either the old computers *or* the old I/O processors taken separately. They were also twice as fast. The use of semiconductor memories was, however, a big leap of faith for NASA: unlike cores, they would have to be constantly powered. Electric power on a spacecraft was still precious; fortunately, the semiconductors did not take much.

NASA was looking forward to getting the faster processors and bigger memories, and had high hopes for them. Previously, memory limitations had forced the agency to trim the software to fit—which meant accepting some limits on what it could do. Now it seemed

that some of the more awkward crew interfaces could be smoothed out, and abandoned functions could be put back into the software. Also, certain parts of the flight load (like the ascent program) had filled all but a few words of the old memory. Fixing certain errors—or adding anything at all—had been nearly impossible. Now there might be room for a few improvements. Sadly, these hopes were overshadowed by grim new necessities.

The *Challenger* disaster changed the use of the new memory. More safety functions had to be added—such as the contingency abort in which the crew could bail out of a crippled Shuttle. The computers would have to fly the spaceplane and keep it steady while the astronauts hit the 'chutes. The new functions required most of the new memory. As a result—even though the entire orbiter fleet would be upgraded to the new machines—most of the software would still be used just as it had been before the upgrade.

Even so, the Shuttle has emerged as the most automated spacecraft ever built. Besides the main computers, there are various specialized processors; one set—used for displays—interacts directly with the crew. The Shuttle is the first spacecraft to have display monitors for output, much like those on Earthbound computers.

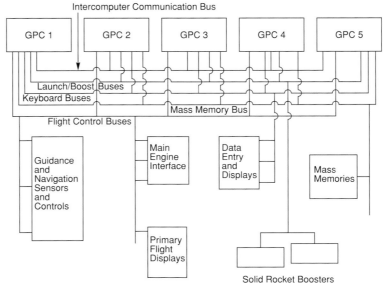

Primary Avionics System bus diagram. Note five general-purpose computers connected to many subsystems.

The displays measure only five by seven inches, and can show 26 lines of 51 characters each. The characters are relatively big (and their number is relatively small); both pilot and mission specialist have to be able to see the data from their stations. There are three screens in the main cockpit between the pilots, and one to left of the mission specialist (facing the cargo bay). On a ship already full of computers, each display also has a special computer (with 8K of memory) to drive it.

Computer-Controlled Engines

As a Shuttle countdown nears zero, close observers can see a cascade of sparks spraying under the three big liquid-propellent main engines mounted at the rear of the orbiter. At about six seconds from launch, pumps send a mist of liquid hydrogen and liquid oxygen into the sparks—one by one, the engines start. In the course of the Shuttle program—several times they have suddenly stopped.

Semiconductors and Volatile Memory
A *semiconductor* device is made of materials (such as silicon) whose ability to conduct electricity can be precisely controlled. A transistor is a semiconductor device. The Shuttle's new semiconductor memories were *volatile*; they would hold data as long as they were powered. To protect against losing any data in case of transient power loss, they had to have a continuous connection to a battery.

If everything is not running perfectly, the engines have to be shut down before the solid rocket

Shuttle main engine test.

boosters are ignited. Once lit, the boosters burn till exhausted, and there is no stopping a flight—even a disastrous one.

This capability of a timely shutdown is one of many design features that made the Shuttle's main engines more complex than any predecessor. As a reusable manned spacecraft, the Shuttle posed some tough design problems. Its engines were no exception. They had to be controllable by a throttle, a rare requirement in previous rocket engines. They had to deliver high thrust with great efficiency. They also had to "know" whether or not they were running right.

The Shuttle main engine's control computer in its test harness.

Overbuilt for ruggedness so they could be reused, the Shuttle engines attained decent (but not completely optimal) fuel efficiency. Their true glory was their control system. All previous rocket engines in NASA's launch vehicles had been "dumb"—completely mechanical—and apart from their size, not too different from the V-2 engine. The Saturn V booster—a complex (but direct) descendant of V-2—had once won an award as the mechanical engineering achievement of the year. Its later cousins, the Shuttle main rocket engines, could have won an award in electrical engineering achievement: they were computer-controlled.

Rocketdyne, the engines' manufacturer, designed each one with a redundant set of two control computers. Originally these were Honeywell HDC-601 machines. Their 16K memories functioned much like core memory, storing bits of data as polarized zones on a wire plated with a material that could be magnetized. The technique was sim-

ilar to that of the old-style audio "wire recorders"—or even tape and disk drives. Plated wire, however, was a lot more rugged than tape and disk. The control computers had to withstand the shock, vibration, and heat of a running rocket engine; they would be bolted right on the combustion chamber.

Having the controllers on the engines meant the engine functions could be handled outside the primary avionics system. The main computers would simply send throttling commands to the rockets, and let the engine controllers worry about making it happen. As a further advantage, each controller could monitor the health of its engine from right on the engine, and send a simple go/no-go signal up the bus to the main computer system. This arrangement would become common in later aircraft and spacecraft.

When the Honeywell computers were replaced by microprocessors, the plated-wire memories were also replaced by 64K semiconductor memories.

What Your Car and the Shuttle Have in Common

Current turbojet engines for aircraft—and auto engines—have controllers that function exactly like those on the Shuttle main engines: they control the mixture of oxygen and fuel being fed to the engine. (Jets and cars can take their oxygen from the air; a rocket has to carry its own *oxidizer* to mix with the fuel.)

Both the Shuttle computers and your auto engine's computer also watch for faults in engine operation. When the "Service Engine Soon" light (or a similar signal) shows on your dashboard, the computer has detected a fault and written it to a memory location that can be read by a device at your service shop. This *fault code* leads the mechanic to the right place to fix. This is a lot better than asking you to describe the exact funny sounds your engine made as your car came to a halt in lane three of the expressway. Somehow it's hard to remember anything but the honks of disgruntled motorists maneuvering around you.

The original Honeywell HDC-601s mounted on the Shuttle engines had another characteristic in common with the early controllers on auto engines: programmers wrote the code for both of them in *assembly language* (a difficult symbolic language that tells a machine exactly what to do, one instruction at a time). This low-level approach was error-prone, hard to check, and difficult to maintain.

Both types of computers are rapidly moving forward in their software technology. On the Shuttle's engines, the Honeywell machines have been replaced by more advanced computers. These use the Motorola 68000 microprocessor chip, and are programmed in C (a higher-level programming language). Automakers are also using such higher-level languages. General Motors, for example, uses "Modula-GM" to program its auto-engine controllers. This is a special version of the powerful Modula programming language used to implement basic software engineering principles.

These were duplicated in each computer, to make them as reliable as the plated wire. All four memories could use three separate power supplies. Once loaded from the mass memory tape, the flight code would be really difficult to lose! The engine controllers had attained the needed reliability.

Software for an Aerospaceplane

The Shuttle's Data Processing System was designed to do everything that needed to be done on a mission: flight control, systems management, and payload deployment. It continued the Apollo tradition; the Apollo computer had a hand (or at least a clock) in nearly every spacecraft system. From the standpoint of computer programming and software maintenance, the complexity of the software for the on-board computers kept increasing. At the time, engineers programmed nearly all embedded computer systems like those on spacecraft, using assembly language. That made the software even more complex and difficult to maintain.

In the Apollo program, the MIT team had made their job somewhat easier by creating a special programming language. Their creation combined the lower-level statements of assembly language into a clearer (and line for line, more powerful) language. It had done well in implementing the special navigation equations needed to go to the moon. The new language was *interpreted* (as most versions of BASIC would be in the home computers yet to come), as opposed to being compiled. It was a little slower, but the gains in clarity offset any performance loss. This became the basic argument that would accompany later decisions on whether to use higher-level programming languages: how inefficient would they be?

The Programmers' Efficiency War: Human Versus Machine

The first computers had only machine code for programming. With the development of the FORTRAN and COBOL languages in the late 1950s, it became possible to write equations or create business forms in near-English statements; these were then sent through *compiler* programs that generated machine code. In those early days, hard-core assembly-

The Shuttle and the two environments it must navigate: space and Earth's atmosphere.

language programmers scoffed at compiler-generated code as being too slow and awkward. It did not take long for them to be silenced; the new high-level languages were providing immediate gains in productivity for the scientific and data-processing communities.

Ten years later, the argument started all over again. As computers became more prevalent in real-time applications, speed gained greater importance; NASA saw some advantages from using the interpreted language in Apollo. Knowing the Shuttle would be much more complex, however, the agency wanted the best of both worlds: a language designed for the specific job, which could still be compiled.

There was resistance from some engineers who were unconvinced that a high-level language could be compiled into machine code good enough for the time constraints of the Shuttle. NASA contracted with Intermetrics, a Boston software company (with many former Instrumentation Lab employees), to create a new language for the Shuttle.

127

The result was HAL/S: an algebraic, special-purpose, high-level language that included many of the constructs a real-time system would need. Dick Parten, NASA's first chief of Shuttle software, set up a contest to convince the doubters. He had a team of really good AP-1 programmers write some routines in assembly language, and had similar routines written in HAL. When executed, the hand-written machine code beat the compiled code—but only by about 15 percent. The compiled code's performance hadn't fallen too far short, especially in view of the expected gains: preparing the compiled code would be simpler; the resulting software would be easier to maintain. HAL was in.

Even with HAL available, however, some of the Shuttle software would still be done in machine code for maximum performance. (The biggest piece—the FCOS, or Flight Computer Operating System—would fill nearly 35K of memory.) Still, most Shuttle software (including all applications programs) would be done in HAL.

Making It Perfect

The Apollo computer, with its unchangeable "core rope" memory, had required great precision in its software. Though the Shuttle's memory did not use core ropes, the software still had to be as nearly perfect as possible before launch. This was (and still is) the goal of all NASA's software contractors.

Even so, the software for the Shuttle was not created all at once—it was built incrementally, over the last 20 years. IBM's Federal division (which had many name changes until Loral bought it in

STS-7 in orbit in 1983. One of orbiter *Challenger's* happier missions.

December 1993) did most of the work for the first ten years of the Shuttle program. In the early 1980s, NASA rebid the contract to prepare flight loads for the various missions.

Originally, the agency wanted the winning contractor to handle two tasks: providing the periodic major releases (revisions) of the software, and also the mission-by-mission collection of programs that performed custom-tailored functions. The basic software would fly mission-to-mission, remaining consistent. Specific payloads, however (like *Galileo* or the Hubble Space Telescope), would require custom coding in the payload software, plus new parameters for the *nav* (navigation) software.

Compilers Versus Interpreters

The lowest-level way to communicate a program to a computer is *machine code.* Unfortunately, it is difficult to write, and worse to maintain. High-level language statements like "IF-THEN-ELSE" replace many lines of machine code with a few lines. There are two ways for the high-level languages to be put in a form the machine can execute: *interpreters* and *compilers.* An interpreter "looks" at one high-level language statement at a time, translates it into machine code, and executes it. Then it goes on to the next statement. A compiler is a software program that takes the *entire* high-level-language application program as input, and generates output in machine code. The computer runs the output of the compiler. It is much faster than interpreters.

A newly-formed organization called RSOC (pronounced *R-sock*)—a specially-created subsidiary of Rockwell—won the contract. It quickly became obvious that the learning curve for maintaining the software was steep; it would cost too much time and money for RSOC to learn everything needed do all the maintenance.

The responsibilities were divided: IBM would do a new release of the software every 18 months. This revised version would contain upgraded features, and fix the highest-priority defects. RSOC would prepare a flight load for each mission; IBM would be responsible for putting it onto the mass memory tape.

Up to this point in the Shuttle program, Rockwell (RSOC's parent company) had done the job of programming the Backup Flight System. NASA thought having a separate software team do the backup would mean less likelihood of duplicating an error. (Recent research by computer scientists seems to indicate this is not entirely true. For instance, if an error in the requirements is not caught before coding, it probably *will* show up in both software loads. Conversely, two different sets of engineers are trying to understand the requirements; they are more likely to *find* such errors.

Super Software

The software development process IBM built for the Shuttle demonstrated a highly advanced level of sophistication and skill. The Software Engineering Institute had described such a level as the fifth (and highest) of its Capability Maturity Model—its way of assessing how well a company or division builds software. This was the first — some still say the *only*—time Level Five had been reached. Of the Model's five levels of key practices needed to build software, most government contracting organizations had been assessed at Level One or Two.

Level One was the least organized—equivalent to the chaotic style of program development most hackers or self-taught programmers might display. Level Two included some strong project-management traits, of the sort that delivered such good software as the Gemini and Apollo programs. But as one Apollo engineer would later remark, "quality was bought with money." A spacecraft with the Shuttle's expected service life of 50 years needed something better. IBM, with NASA's constant support, delivered something

"I Can't Do That, Dave—"
HAL, Language of the Shuttle Orbiter Computers

HAL is like other programming languages used for engineering, such as FORTRAN. It has the usual features, such as *looping*, *if-then-else*, and *assignment statements*. What makes it different is that it can implement *vector mathematics* (necessary for space flight) efficiently.

HAL was one of the first languages to include the constructs for *multiprogramming* so useful in real-time systems. Not until the Ada language—nearly a decade later—did another computer tongue speak *tasking* (a *task* is another name for a program) so clearly.

Another neat feature of HAL is that its listings come out three-lines-to-a-statement; this allows enough space between lines that subscripts and superscripts are right where they would be if an engineer wrote the equations by hand. The listings are easier to read.

HAL can schedule tasks in real time using either a *synchronous* or an *asynchronous* approach. *Aperiodic* tasks (another word for asynchronous—things that happen unexpectedly) are handled like this:

SCHEDULE ERROR4 ON RUPT2
PRIORITY15

The statement means that when an interrupt named "RUPT2" occurs, a recovery routine named "ERROR4" ought to be scheduled, and

given a priority level "15." HAL manages the various interrupts by assigning such responses to them. Things that need immediate attention get a high priority setting; lower-priority settings mean slower responses. *Periodic* tasks (those happening at known or desired intervals, like engine monitoring) are scheduled like this:

SCHEDULE ME1 PRIORITY(10), REPEAT EVERY 6./90.

This statement sets the execution of task "ME1" at fifteen times a second, assuming nothing of higher priority is locking up the processor.

The origin of the language's name has been the subject of much speculation over the years. Some say it is named after the rebellious computer in the film *2001: A Space Odyssey*. Many think that particular HAL got its name from the letters of the alphabet that precede (respectively) "I," "B," and "M." Other guesses include the acronyms Houston Aerospace Language and High-level Aerospace Language. Intermetrics folks insist that HAL is named after J. Halcomb Laning, a brilliant language engineer who made significant contributions to compilers and higher-order languages in general. But then, why the acronym-style spelling, all in capital letters? Is it computerese from people used to using all caps on punched cards? Does the tribute to Halcomb Laning reflect a desire to keep *2001* creator Stanley Kubrick (a contentious and eccentric fellow) off their backs? It doesn't really matter. Mr. Laning certainly deserves the honor.

better, and has constantly improved it—the hallmark of a Level Five organization.

When IBM began work on the Shuttle software, it did development in what was known as the "Software Development Laboratory." The actual

computers in the SDL were IBM mainframes with huge memories and even bigger disk drives (at least for those days!). NASA kept its computers in a building at the Johnson Space Center in Houston; IBM set up shop across NASA Road

One. Programmers wrote their code and had it punched on cards. A courier would run the card decks back and forth from the computer rooms to the IBM building.

During the "laboratory" years, all the supporting equipment and software gradually came on line. An interface to some AP-101s like NASA's let IBM load flight code into the actual target hardware (the machines that would run the finished software). Big simulation programs in the mainframes allowed further testing. A terrific version-control system kept track of the software as it took shape. This approach was not only wise, but much needed—the goal was to keep everything built in case it was needed again, or to do a postmortem on a flight load.

The first four flights were the Shuttle test program; after that, the Space Transportation System would be operational. During that time, the SDL made a corresponding transition: the "Laboratory" became the Software Production Facility (SPF). Building the software had become less experimental and more routine.

Making Software "Human-Rated"

IBM software engineers used five major layers of tests to verify the software's readiness to fly. The first of these layers involved inspecting the code. Ordinarily, when big, modular programs like the Shuttle's are built, individual units of the software would be tested before they were integrated with others. Their size could range from a few lines to a few hundred lines of HAL/S code. Many programmers used "wring-it-out" testing on their code, but IBM dropped unit tests in favor of inspections: the developer and two or three other engineers would go over the code line by line. These inspections actually found more errors than unit tests ever did.

The software was then integrated and tested as a system (the second layer of tests), including some testing of mission-specific data (third layer). An independent verification team would also do general system testing and certify the software's parameters (fourth layer). Finally, a mission load

Getting Out of Trouble: Aborts and More Aborts

One reason the ascent load is so big is the variety of aborts it contains. These emergency programs must be kept loaded in the computers so there will be no delay in activating them if a problem appears. Some of them are very dangerous and complex. Others are fairly benign, and have actually been used.

The most likely problem on ascent is the failure of one or more of the three main engines. If such a failure occurs late in the ascent, the fuel for the failed engine is rerouted to the other good engines, and a slightly longer burn takes place. This happened on a flight in 1985 with no trouble. By the time the engine went out, the solids had been jettisoned—and weight of the fuel remaining in the tanks was small enough for two-engine operation. That is a form of "Abort to Orbit," or ATO. Other

single-engine-out aborts can send the Shuttle across the Atlantic to landing sites in Spain or Morocco. If complete primary computer failure (or some other catastrophic subsystem failure) occurs past the point where transatlantic aborts could be done, AOA ("Abort Once Around") is used. The Shuttle orbits once, and immediately performs re-entry.

Two of the most dangerous aborts are "Return to Launch Site" (RTLS) and a contingency abort added after the *Challenger* disaster to enable the crew to use the new escape system. RTLS enables a crew with an early engine failure to throw off the solids, and fly a reverse course under power to Kennedy Space Center. This is a tough aerobatic maneuver, full of aerodynamic difficulties. Also, the escape system needs the orbiter to be flying essentially "straight and level," so the new contingency abort is no flight-control picnic, either.

would be delivered to the Shuttle Mission Simulators and the Shuttle Avionics Integration Laboratory for the crew to use in training (fifth layer). Sometimes the astronauts themselves would find errors, but not very often.

Even with all this effort, some errors could slip through. If these were not life- or mission-threatening, however, they could be "waived" until the next major release of software. For example, some years ago it was possible to get an unexpected pitch command if a really rare and strange set of circumstances came about. As long as the crew knew what those circumstances were—and simply avoided them—the uncommanded pitch did not occur. Such known bugs would be noted in a document called "OPS Notes and Waivers." To find out what an OPS is, we'll have to look at the architecture of the software.

OPS, SPECs, and DISPs: Structure of the Shuttle Software

As it exists today, the Shuttle software retains its original architecture. Since an entire load of flight software cannot fit into the Shuttle's Data

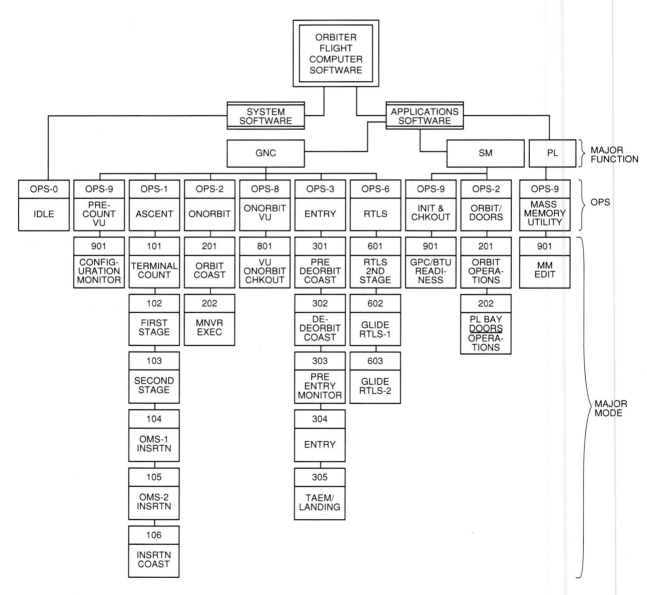

Structure chart of the Shuttle's on-board software.

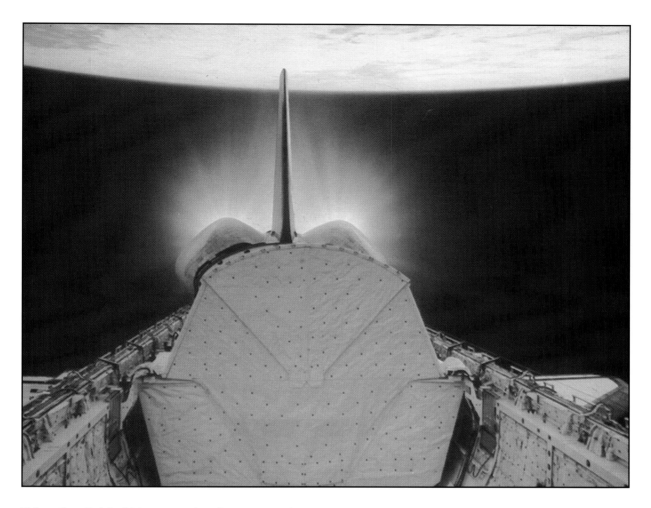

Firing the Orbital Maneuvering System engines.

Processing System all at once, it has to be logically divided. The software has both vertical and horizontal layers, as you can see in the chart.

The highest level on the structure chart contains just two items: *systems* software and *applications* software. That division ought to be familiar to any computer user; the former type runs the computer system, the latter type handles specific tasks. For example: DOS and Microsoft Word, UNIX and a graphics program, VM/CMS and CICS. In this case, the systems software is FCOS; the flight-control software is an application.

The applications software is further separated into three major *functions*: guidance and navigation, systems management, and payload. Each function has one or more "OPS," or *operational sequences* (a series of commands that will perform a task).

Each sequence is usually assigned to a specific phase of the mission—and each is the entire applications load of a particular computer.

In turn, each operational sequence is made up of one or more major *modes*. These are collections of programs that run in real time, performing specific tasks. Each major mode has a distinct set of running tasks. Transition from one major mode to another is automatic.

As an example of how the OPS and major modes work, let's take OPS-1, the ascent program. It is loaded into the redundant set of computers at about T-minus-30 minutes in the countdown. It boots up in Major Mode 101, "Terminal Count." MM101 handles the final parts of the countdown in sequence, until a few seconds before launch—when Major Mode 102 takes over. MM102

133

```
 2011/000/                    GPC MEMORY      2 137/22:10:18
                                              000/08:10:18
MEM/BUS CONFIG                READ/WRITE GNC
1 CONFIG   _3 (G3)             DATA 20* BIT SET 22   SEQ ID 24
2 GPC   1  2  3  4  0          CODE 21  BIT RST 23   WRITE 25
                              26 ENG UNITS ___       HEX 27*
  STRING  1   7  1             ADD ID   DESIRED   ACTUAL
          2   8  2            28 _____   29 ____
          3   9  3            30         31
          4  10  4            32         33
  PL 1/2     11  0            34         35
                             36         37
  CRT     1  12  1            38         39
          2  13  2
          3  14  0            _____
          4  15  0            MEMORY DUMP
                              40 START ID    _____
  LAUNCH  1  16  0            41 NUMBER WDS  _____
          2  17  0            42 WDS/FRAME   ___
  MM      1  18  1            DUMP 43
          2  19  3
                              44 DOWNLIST GPC 1


          I/O ERROR          CRT2         4      09:59:58()
```

The GPC Memory Display, one of many SPEC displays available to the crews.

ignites the liquid engines, makes certain they are running correctly, then ignites the solid rocket boosters. Once the Shuttle is off the pad and climbing, MM102 performs the roll program that turns the Shuttle onto its back for flight through the atmosphere. After about two minutes, the solids burn out and are jettisoned.

At that point, Major Mode 103, "Second Stage," takes over. The main difference between MM102 and MM103 is that there are now no solid rockets to control. The "Boost Throttling Task" remains in the set of tasks running in MM103, still controlling the thrust of the liquid-propellant engines. The task that gimbaled the solid-rocket nozzles for steering is put to sleep, no longer needed.

After the liquid-propellant engines burn out, the computers make the transition to MM104—firing the Orbital Maneuvering System engines (mounted in two humps at the tail of the orbiter) for the first time in the mission. That first burn actually inserts the Shuttle into the correct orbit. Then MM105 takes over, controlling a second firing that refines the orbit to match mission parameters.

The final major mode of OPS-1 is MM106, "Insertion Coast." That one simply maintains navigation data and spacecraft systems until the pilots can reconfigure the Data Processing System for on-orbit operations.

The crew can interact with the software using screens called SPEC and DISP. Each OPS has a set of SPEC and DISP displays, and there are some that can be called up in any OPS. There is also a set of OPS screens, unique to each OPS. As the spacecraft progresses through the mission phase supported by a particular OPS, its screens show information to the crew automatically. For example, OPS-3 has a re-entry trajectory display that shows the actual and projected flight track (a Shuttle icon moves to show relative position). It also has a bar display that shows altitude and velocity.

One DISP shows data about the auxiliary power unit and hydraulic system. Because these devices have a general importance, this DISP is available in more than one OPS. Another screen that has an importance to the whole system—and so is in all OPS—is a SPEC: the GPC MEMORY display. This system-level interface to the computers sets up which keyboard and which monitor are connected to which of the five general-purpose computers. GPC MEMORY also can be used to assign *strings* (that is, specify configurations of buses and machines). Its other functions include actually changing the data in memory, initiating a memory dump, or transmitting data directly from a computer to the ground.

Flying the Shuttle Computers

How does an astronaut use such a display? Let's say Mission Control asks STS-34 Mission Commander Don Williams to downlink some critical data so that the ground-based computers can check it. He needs to use the functions located under the heading "MEMORY DUMP" (in the lower right portion of the GPC MEMORY screen shown here). How does he get to those functions? The keyboard on the Shuttle (as shown in the next drawing) is fairly limited, though not so limited as that of Apollo. Note that it is not a full alphanumeric keyboard. The "A" through "F" keys are not there for any alphabetic reason, but rather because this is a *hexadecimal* (base-16) keyboard.

When the Primary Nav Blacked Out

Turning off on-board computers has backfired at least once. On STS-8, the crew was beginning to get ready for re-entry and commander John Young fired the nose attitude control jets. Something akin to a backfire made the nose jump like the Shuttle had been hit with an uppercut. The sudden jolt caused some solder to fall off a circuit board in the primary nav computer and it shorted out, leaving the spacecraft without the current state vector and no way to navigate. The crew switched on the other computers, but had to stay in orbit longer while the updated navigation data was transmitted and then checked bit-for-bit. Once this was done, *Columbia* made a normal landing. The shorted computer had come back on line, but it failed again just as the main wheels touched down.

Some of the special function keys around the outside of the keyboard may seem familiar to you from the Apollo piloting information earlier in the book. OPS and SPEC are fairly obvious: to load a chosen OPS from the mass memory tape, the commander would press OPS, a number, and EXEC. SPEC works the same way.

Right now Williams needs to get that memory dump to Mission Control. He presses ITEM 40 on the keyboard: a star (like the one currently to the right of item 20) appears next to 40 START ID, and the five underscores to the right appear highlighted on the display. Williams then enters the hexadecimal address of the starting data word in the desired dump. He then presses EXEC. Pressing ITEM 41 enables him to enter the number of computer words to be sent in the dump. He can then go directly to ITEM 43 to initiate the dump—but he should also indicate (in ITEM 44) which computer is the current one, so the correct memory can be downlinked. This procedure represents a routine interaction with the software. Following STS-34 through its mission will illustrate how a typical astronaut crew flies the computers.

Piloting the Processors

At the specified point in the countdown for STS-34, Commander Don Williams and Pilot Mike McCulley load OPS-1 into all four primary avionics-system computers. The Backup Flight System is already loaded and running in the fifth

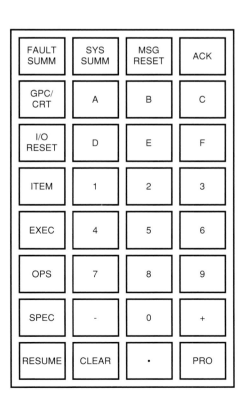

The hexadecimal keyboard used on the Shuttle.

Forward view of the orbiter cockpit, commander's seat to the left, pilot to the right. The three monitors and two keyboards are visible between their seats. The failure matrix is the black square above the window on the commander's side; the switches directly above it control power to the individual computers.

On the three forward display monitors, the crew has selected the ascent displays of both the Primary and Backup Flight Systems. Ironically, they prefer the backup's ascent trajectories: one screen clearly shows the point where the RTLS abort will no longer be possible, and also the point where some orbit can be achieved. The third display monitor is for system management.

During the ascent, Franklin Chang-Diaz, the Senior Mission Specialist on board, has the job of watching the computer-failure matrix over Don Williams' shoulder. The matrix is difficult for the commander to see from inside his helmet. Chang-Diaz has a good, but distant, view of the control panels from his seat.

After the OMS engines make the second burn (fine-tuning the orbital insertion), OPS-1 has switched into "On-orbit Coast," its final major mode. Now it is time to reconfigure the Shuttle for on-orbit operations. It is very important to get the cargo bay doors open quickly. Inside the doors are big radiators that help keep the temperature balanced inside the crew compartment. The sensitive avionics are all air-cooled, and the spacecraft has had to hold in all its internal heat since being buttoned up hours earlier.

machine. Major Mode 101 finishes its work; *Atlantis* leaps from the launch pad into a long climb through the atmosphere, under the control of MM102.

To open the doors, one of the computers must run the second operating sequence (OPS-2) of the "System Management" major function. Williams and McCulley use the "GPC MEMORY" SPEC screen to assign the computers, one by one, to appropriate keyboards and displays. Then they load OPS-2 of the "Guidance and Navigation" major function into two of the Primary Flight System's computers. These two are now the redundant set.

Orbital Mechanics Terms

For the English majors in the audience, yes, *in orbit* is a better use of language than *on orbit*, but the latter is NASA-ese for the same thing: moving forward so fast that the Earth curves away under a spacecraft the same distance the spacecraft falls toward it. So, though actually falling along its orbit, the spacecraft never reaches the ground. Hence the related term, "free fall."

The *state vector* is the position, velocity, acceleration, and direction of motion of a spacecraft. From that data future maneuvering can be planned.

OPS-2 has just a pair of major modes: "On-orbit Coast" (to gracefully accept the state vector and other important data from OPS-1) and a cluster of software modules for maneuvering. A third computer is loaded with the "Systems Management" and "Payload Deployment" software—the "other" OPS-2. The cargo doors can now be opened.

The fourth of the primary computers is loaded with the OPS-3 entry software—essential for getting home—in case a later failure prevents loading from the tape (or simply to have ready for any emergency). Once loaded, usually the fourth computer is shut down along with the backup system. Williams has to reach a set of power switches over his left shoulder to turn the machines off. This procedure, called "freeze drying," saves electrical

power. When a particularly amp-sucking payload is on board (such as the European Space Agency's Spacelab module, with its many experiments), only one guidance-and-nav computer is up. Three are turned off.

The first major task on STS-34 is the deployment of *Galileo* and its Inertial Upper Stage rocket booster. Chang-Diaz calls up the payload display on his screen (mounted in a panel to his left) as he faces the cargo bay windows looking out on the interplanetary explorer. This particular SPEC display enables the crew to check out and arm the booster, warm up the spacecraft, and launch it from the cargo bay. (This SPEC, specific to STS-34, is an example of a custom piece of software.)

Just over six hours after launch (at 7:15 p.m. EDT on October 18, 1989), *Galileo* leaves the cargo bay. Its rocket booster fires perfectly; at 8:20 p.m., it enters a transfer orbit for Venus—the first planet scheduled to provide a gravity boost. *Galileo* begins its complicated, indirect, six-year flight to Jupiter.

For the rest of the on-orbit phase of the STS-34 mission, the computers are used mainly to manage various spacecraft systems. There are some small problems with a flash evaporator and the heater in an auxiliary power unit. A DISP screen shows data on these glitches as needed.

The morning of October 23, 1989 dawns reasonably calm at the Edwards Air Force Base landing site, but the forecast is for increasing wind in the afternoon, so the landing time is moved up a

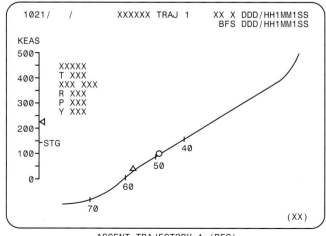

Ascent trajectory display from the Backup Flight System, commonly used by the astronauts during the climb to orbit.

BFS GNC — "SPEC" DISPLAYS

XXXX/050/			HORIZ SIT			XX X DDD/HH1MM1SS		
						BFS DDD/HH1MM1SS		
	ALTM					NAV DELTA		
	9 XX.XX					ΔX 10		
						XXXXXXX		
RWY CTR X						ΔY 11		
XXXXX 3X						XXXXXXX		
XXXXX 4X						ΔZ 12		
XXXXX 5X						XXXXXXX		
						ΔX 13		
TAEM T6T						XXXXX		
GIN X XXXX 6						ΔY 14		
HSI X						XXXXX		
XEP 7						ΔZ 15		
AIM XXXX 8						XXXXX		
						LOAD 16		
	PASS/BFS		SV XFFR 17X		18 ΔT	XXX.XX		

NAV	RESID	RATIO	AUT	INI	FOR	TAC1 XXX	TAC2 XXX	TAC3 XXX
TAC AZ	±XX.XX	XX.XS	19X	20X	21X	±XXX.XXS	±XXX.XXS	±XXX.XXS
RNG	±XX.XX	XX.XS				±XXX.XXS	±XXX.XXS	±XXX.XXS
						DES 31X	DES 32X	DES 33X
DRAG H	±XXXXX	XX.XS	22X	23X	24X			
ADTA H	±XXXXX	XX.XS	25X	26X	27X			
ADTA TO G1C			28X	29X	30X			

HORIZONTAL SITUATION (BFS)

Backup Flight System entry display, showing the horizontal situation. The Shuttle icon is in lower center (above the letters "SV"). The dots show the preferred path: to enter the turn on the edge of the circle, and steer right to the line that represents the extension of the runway centerline.

couple of orbits (about three hours). The cargo doors close; Williams and McCulley configure the four primary computers with OPS-3. Williams turns on the power to the Backup Flight System computer. Once the redundancy-management software is in sync, descent procedures begin.

The pilots have a convenient count-down clock for setting alarms and engine-firing times: the second line of the "mission elapsed time" display in the upper-right-hand corner of all screens.

Atmospheric entry is actually a busier time for the pilots than ascent, just as preparing for landing is in conventional aircraft. There is added pressure in a Shuttle landing: no go-arounds (gliding in to a landing means no second chances). If the astronaut crews

PCs in Space

When the Shuttle began flying in 1981, it was not possible to stay in continuous radio contact with it. There were gaps in the string of ground stations that had banded the Earth since the Project Mercury days. The pilot on the first Shuttle mission, Bob Crippen, had the idea of using a programmable calculator (remember, this was in the days before laptops) to figure out the time the next ground station ought to be in range. NASA bought a Hewlett-Packard HP-41C, and it became the personal computer of the crew for a few years. If it had been considered as a flight article, it would have cost millions of dollars to make it "space-rated." As "personal property" (each crew member gets a few pounds' allowance), it did not have to be

certified. The astronauts wrote the programs themselves and used it in flight.

The HP-41C also had a primitive timing program, allowing astronauts to calculate the time-to-retrofire for certain landing sites (a hedge against loss of the state vector at an embarrassing time). Capitalizing on this creative solution, Hewlett-Packard took out full-page ads in the *Wall Street Journal* that touted the calculator as (more or less) the ultimate backup computer. Once portable computers came available, however, the crews "upgraded" to a Grid Systems Compass computer with some graphics capability. They could use it like the display in Mission Control, to show the ground track (the orbiter's path across the planetary surface) and the upcoming stations.

Returning from space, the Shuttle performs a "dead-stick" (unpowered) landing in which computers play an important controlling role.

make a mistake in the important area of energy management, they could land short of the run- way—and embarrass everyone involved (the flat lakebed at Edwards still offers ample room to land).

It is possible for the com- puters to fly the spacecraft all the way to the runway for a landing. That's how the Soviet shuttle *Buran* flew in its only test. As you can imagine, a typical astronaut pilot wants none of that (there's a "no- Spam-in-a-can" tradition to uphold). So they usually take over at 40,000 feet and fly manually, though they get a lot of help from the software.

There are five "entry-trajectory" displays: four are "vertical situation" screens, and one is a "horizon-

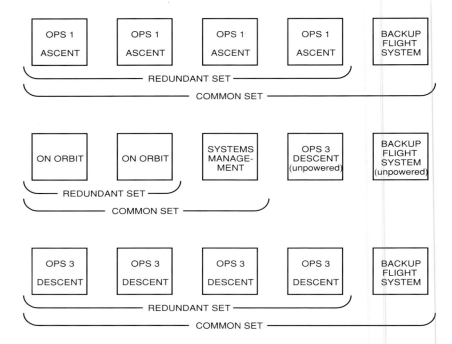

During a typical flight, the Shuttle's five general-purpose computers are configured in at least these three ways. The top arrangement is for ascent to orbit, the center one for on-orbit operations, and the bottom configuration for return to Earth.

tal situation" screen. The trajectory displays show flight-path-versus-planned-path, a history of the entry track, and some data on velocity and alti- tude. The vertical situation shows attitude data (such as nose position relative to "nominal," that is, ideal), and vital flight-control data such as the angle- of-deflection of ailerons, rudder, and trim tabs. The horizontal situation is a complex screen showing change-in- velocity in three axes, as well as a plot of the Shuttle's position relative to a circle that has the landing run- way as a tangent.

To reach a point where the runway is "made" (successfully approached), the Shuttle must bleed off the energy of its forward momentum—and, gradually, of its lift—in a very precise

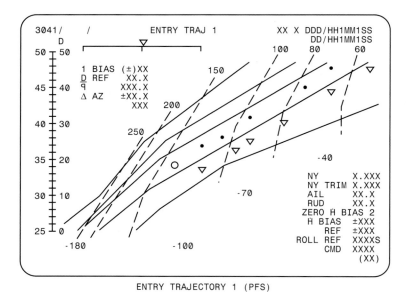

ENTRY TRAJECTORY 1 (PFS)

Primary Flight System entry display, showing the vertical situation of the orbiter.

Jim Tomayko in the cockpit of his Aero Club's Piper Warrior, which can do something the Shuttle can't: a go-around.

It's a Brick

Having practiced flying landings to Runway 23 at Edwards in a Shuttle simulator, I can appreciate what some pilots say of the Shuttle's flying characteristics: "It's a brick." Although my opinions are those of a relatively inexperienced private pilot, here goes: once down to a certain altitude, things are easier, since the computer has done most of the energy management. Taking over at about 20,000 feet, finishing the turn to line up with the runway, and then managing the remaining part of the approach is helped a lot by the information given in a typical heads-up display (such as altitude, airspeed, and descent rate).

If you make a mistake lining up, however, it's tough to center on the runway. In a powered aircraft, you can (sort of) just fly to the centerline, and if there is any crosswind, bank into a slip (raise a wing slightly, to lose a little lift) to hold position. The Shuttle does not respond as easily to slips. This may be because it flies like a *lifting body* (the whole body is designed to lift, not just the wing); banking the wing may not reduce lift as much as on a conventional aircraft. I did make the runway, once in every three tries or so. Luckily Edwards has a lot of open space!

Sitting through a computer-controlled landing is unnerving. The descent rate is quite high, and the computer does not make any attitude mistakes that might cause you to "balloon" on landing, or overshoot the runway. You'd swear you were coming straight down, and the landing gear must plunge into the ground—but just as you get there, the nose comes up, the last bit of energy bleeds away, and the wheels kiss the concrete. Of course, it could all be a conspiracy of the simulator computers not to make the Shuttle computers look bad . . .

—*Jim Tomayko*

way. The landing approach is a turn to place the Shuttle on the circle that shows on the horizontal situation screen; the resulting bank to line up with the runway takes away the excess energy.

Atlantis touches down at 12:33 p.m. EDT, ending the short five-day mission of the Shuttle and its crew. Meanwhile, *Galileo* continues a voyage to

Jupiter that will take over 400 times longer, followed by years of orbiting in the midst of the complex Jovian system of satellites. *Galileo's* journey—and that of its computers—only *begins* in 1989.

Twin Bases on the Red Planet

9

When the Shuttle deployed the *Galileo* spacecraft in 1989, it was launching the most recent of a highly successful series of space probes. These (as you may remember from Chapter 4) had been built by the Jet Propulsion Laboratory (JPL). Early probes were strictly *flybys* of Venus and Mars; they passed the planet and went on. Later missions orbited both Mars and Venus, and zipped past Mercury, Jupiter, Saturn, Uranus, and Neptune—every planet in the Solar System except Pluto. The probes' missions lasted for years; some launched in the 1970s are still in progress. Though they never return, some continue to explore.

The great scientific gains made by these probes have kept alive the manned-versus-unmanned-spacecraft debate. Obviously, the loss of *Mars Observer* in 1993 is less keenly felt than the loss of *Challenger* in 1986. One of the chief benefits of having a human crew aboard— overriding failed hardware—entails real risk, and sometimes comes at a great cost. Without living things on board, a spacecraft needs no life-support equipment, and safety is not as big an issue. These factors help make unmanned spacecraft cheaper.

A mosaic of 102 Viking orbiter images of Mars, processed to show how the Red Planet might look to an approaching spacecraft.

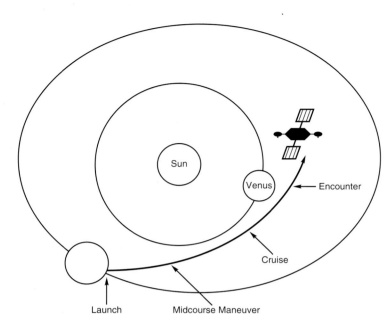

Typical flyby of Venus, showing the four mission phases.

On the other hand, humans have a strong desire to experience things in person, rather than vicariously. (This may change as more of what now passes for reality becomes "virtual reality," and people get used to living in it. Then being "plugged in" to a robot probe wandering on the surface of Ganymede might not seem much less exciting than being there yourself.) The debate continues; neither "manned" nor "unmanned" advocates seem willing to be left out of space flight.

Since humans are not aboard the space probes, however, the probes have to be as resilient as possible. They have to be able to detect and recover from failures. To achieve this, the space probes carry ever-more-sophisticated computer systems on board.

Computers That Weren't: Mariners Fly to Mars, Venus, and Mercury

Deep-space missions all combine the same four components: *launch, midcourse maneuver, cruise,*

and *encounter*. In the case of a single-planet flyby, everything happens just once. The probe is injected into an intercept orbit by the booster vehicle. Within a few hours or days, its course is corrected as needed. The "mid" in "midcourse maneuver" is actually more like "early cruise"; it has to come early and be accurate. If flight controllers were to wait until the middle of the cruise to change the orbit, a small navigational error could grow into a detour of many miles. Such an error would be too large for the limited maneuvering engines of an unmanned spacecraft to handle.

In the absence of warp drive or other science-fiction propulsion systems, cruise is always the longest mission phase. It ranges from a couple of days to the moon to over a decade for the outer planets. Finally, the encounter is either a few hours (in the case of a flyby), or months and years (in the case of an orbiter like *Galileo*). For multiple-flyby missions—like the journey of Voyager I past four planets—each probe repeats the steps of course correction, cruise, and encounter for each planet.

During each of these mission phases, the spacecraft itself has to perform various gyrations. Just after launch, it must deploy its antennas and instruments as it reaches space. It has to change attitude, and time the thruster firings in the midcourse maneuver. For the entire cruise period, it monitors its miniature on-board science laboratory, and keeps the antenna pointed to the ground stations so data can be downlinked to them. During encounter with a planet, it has pictures to take and sensors to point. All these things occur in a fixed sequence of operations.

The Ranger VII spacecraft.

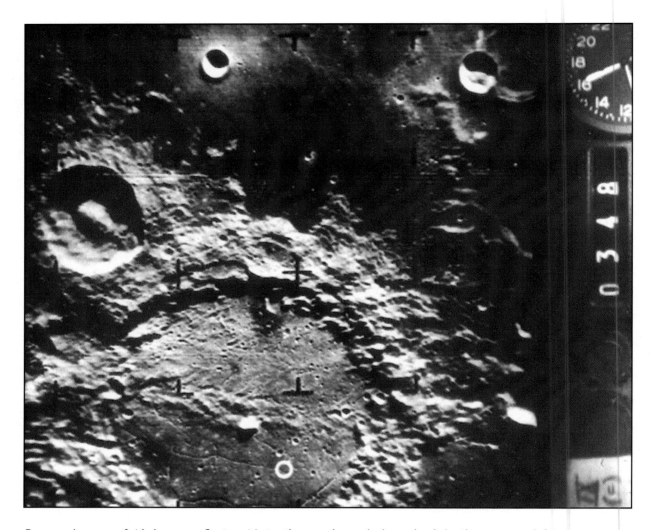

Ranger image of Alphonsus Crater. Note the analog mission clock in the upper right.

At launch, a probe's computer contains a simple version of the entire mission sequence. As each phase of the mission is successfully completed, controllers can make updates if the probe still has receivers working. If it does not, then it executes the current version of the sequence. As long as the transmitters and receivers work, all is not lost.

The first probes built by JPL had short lives. Each of the Ranger spacecraft, for example—when they worked—covered the distance between the Earth and moon in a couple of days, and ended the mission with impact on the moon. The Rangers could be given commands directly from the Deep Space Network, since the longest speed-of-light delay was still under a second. In case of receiver failure, a Ranger could still accomplish part of its mission; it carried a "Central Computer and Sequencer" that kept track of when the cameras and other instruments should be turned on. The only catch was that the computer only computed time—basically it counted "one thousand one, one thousand two," and so on, from a time before launch until its lunar crash-landing. Artificial intelligence it wasn't; it could do *nothing* to correct faults. Therefore, if the Ranger missed the moon by its customary 20,000 miles, the computer would take photos of whatever was in front of it at the time—which was also nothing.

The first Mariner missions to the planets were more sophisticated. Their version of the Central

Computer and Sequencer could hold quantitative commands ("do *this* much of *that* for exactly *so long*") for later use, and would still initiate the commands and keep count of time. The clock was an oscillator, just like computer clocks today.

Missions to Venus and Mars meant long delays in radio reception, since messages had to cover vast distances at the mere speed of light (no subspace transmitters here). Therefore the Sequencer on the spacecraft was itself the primary source of commands. A Mariner would take off with the basic mission sequence essentially hard-wired in. Controllers could send up some numerical time-values for duration—such as "fire motors for 12 seconds"—or as alarm clocks (such as "begin photos in 134 hours, 15 minutes, 18 seconds"), and

the probe could store these in a limited memory. In light of the versatility and power of later computers, it seems unfortunate that these devices had the word "computer" in their name. It gave the wrong idea about their true capabilities.

When the time came to plan for the Mariner Mars 1969 mission, JPL wanted to make the on-board sequencer more functional and powerful. One reason was to gain some flexibility during encounter. Two spacecraft would be launched days apart. If the first one found something interesting, like a Martian shopping mall, it would be nice to be able to redirect the second spacecraft to make another pass. JPL designed (and Motorola built) a 26-pound, programmable version of the Central Computer and Sequencer.

Mariner 4 spacecraft sent to Mars.

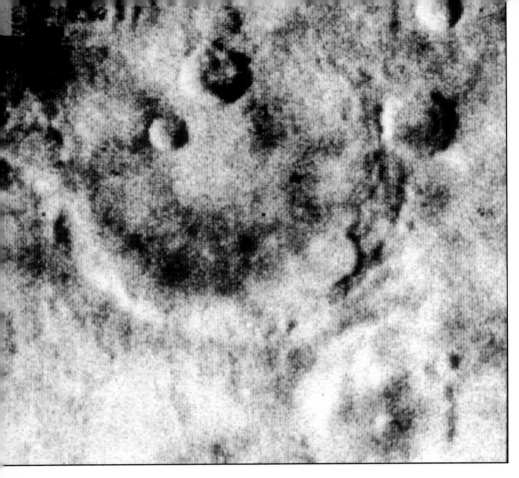

One of the first close-ups of Mars—the image that broke a thousand hearts (who hoped to see signs of civilization, instead of a cratered surface like the moon).

New Features, New Techniques

The new Mariner version of the Sequencer had an "enormous" amount of memory—128 words, each 22 bits long—and little else. It had no accumulator, nor most of the other registers found in a conventional computer's central processing unit. It did have adding and subtracting circuits, but all working data was stored in the main memory, leaving little room for elaborate programs.

Instruction words, event words, and data words were the three types of 22-bit words in the Sequencer. Sixteen instructions were available. The first four bits of an instruction word were the actual instruction. The next nine bits gave the address of the data to be manipulated; the last nine bits specified the address to read next.

For example, suppose the instruction told the machine to "decrement minutes and jump." The Sequencer would find the data (the value of *minutes*) stored in the event word at the first specified address; upon finding that value, it would *subtract one* ("decrement"). If the result was greater than zero, the Sequencer would "jump"—find the instruction at the address stored in the last nine bits of the instruction word, and execute it. If the result was zero, the data in the event word had "zeroed out." In that case, the Sequencer would go to the address specified in the last nine bits of the event word it had just visited.

The event words were "alarm clocks." Decrement instructions counted them down until they hit zero, then something different would happen. The first seven locations in the spacecraft's 128-word memory stored a simple "executive" program, executed each hour. If one of the event words that counted hours zeroed out, a simple program would start: first decrementing minutes, then counting down seconds, and then commanding the spacecraft to turn on cameras or send messages. These alarm clocks occupied nearly the entire memory.

When they launched, each Mariner spacecraft of the 1969 Mars mission held a bare-bones version of the total mission in its 128-word memory. After they attained Mars-intercept orbit, however, they no longer needed the code for the completed phases of launch and midcourse maneuver. Therefore JPL replaced it, painfully uplinking a more robust cruise-and-encounter program to the probes. As the distance increased, the transmission rate slowed to a crawl: it took nearly two minutes to replace one word. But changing this memory load in flight enabled spacecraft to get by with much smaller memories than would otherwise be possible. It was the beginning of a JPL tradition.

Bit-tweakers among the readers will have figured out by now that with nine-bit addresses, it was possible to give the spacecraft a memory larger than 128 words—*without changing the logic hardware in any way*. For the Mariner Mars 1971 orbiter missions, the memory was indeed expanded to 512 words. The Sequencer got a lot of work on those flights. Since the encounter phase was actually to orbit Mars, the flow of science data and photographic imaging would last longer than on any previous planetary mission.

Orbit Changes at No Cost

The midcourse-maneuver phase of a mission is often accomplished using the help of a planet's gravity, rather than rocket fuel. Mariner X had its orbit changed during its Venus flyby to put it on a trajectory for Mercury. Later, both Mercury's gravity and that of the Sun helped position the probe for later encounters. Ground-based navigation programs figure out where to aim the spacecraft to take advantage of gravity assistance.

When the Mariners arrived, Mars was in the midst of one of its biggest dust storms; the entire planet was obscured. Imaging had to wait for weeks, so

Mariner Mars spacecraft, 1969.

Improved imagery from the Mariner VI flyby of Mars in 1969.

the Deep Space Network kept busy updating the alarm clocks in the spacecraft memory. When the imaging finally began, it returned the first really high-quality photos of the Red Planet, including a spectacular shot of Nix Olympica—the largest volcano in the Solar System—poking above the clouds.

Only the Mariner Mars 1971 missions would make incomplete use of the available on-board memory in their computers. Both spacecraft did manage to use over 400 words each, but never came close to the memory overflows so characteristic of spacecraft computing. On the next Mariner flight, things went back to "normal."

Mariner Venus-Mercury 1973, the tenth and final Mariner mission, had a busy schedule. Several mis-

sion phases were practically missions in themselves: a Venus gravity-assist flyby, a triple flyby of Mercury, and some en-route imaging of the Earth and moon. With what it could carry in its memory at launch, the Central Computer and Sequencer had nowhere near enough horsepower to run a full-blown version of each "sub-"mission.

The computing strategy called for a series of program changes in flight. Each would give the probe a detailed program for its next mission phase, and fill the remaining memory with a no-frills program for the rest of the mission. One of the "alarm clocks" was always set to go off if a command had not been received from Earth in a specified period of time. That way, if the receivers failed, the flight would still go on. As the probe completed each mission phase, JPL sent up a new program detailing the next phase, along with the no-frills backup.

The amount of reprogramming would, obviously, be extensive. In previous missions, the programs had been built by hand. For Mariner X, a command-generator program was available to speed things up. This upgraded version of COMGEN had first been used during the Mars orbiter missions. Its task was to help format commands, and then to transmit them automatically to computers in the Deep Space Network stations. From there, the commands were uplinked to the probes. The program got a lot of work during Mariner 10.

Back to Mars: Twin Orbiters, Landers, and Computers

Sent to Mars in the mid-1970s, the Viking dual spacecraft—two sets of orbiters and landers—

made the first use of "real" computers on unmanned deep-space probes. They achieved Martian orbit in the summer of 1976; the landers descended to the surface, while the orbiters stayed aloft mapping and doing scientific sensing. The landers' chief mission was to carry a sophisticated chemical laboratory that would analyze soil samples and figure out if there was life on Mars.

Since these were extended, highly complex missions—with a need for flexibility (and a decent budget)—computers were designed in from the outset. But the stories of the orbiter computers and the lander computers would be quite different from each other.

The Viking mission was the first and only time that Langley Research Center in Virginia had overall responsibility for a deep-space mission. The Langley program managers chose JPL to build the

The Mariner Mars mission of 1971.

orbiters, but a commercial contractor—Martin Marietta—to build the landers. JPL decided for this mission to replace the Sequencer with the Command Computer Subsystem, or *CCS*. The CCS consisted of a dual-redundant string of components: two computers, two power supplies, two output units, etc. This marked an advance in reliability. The old Central Computer and Sequencer had carried a hard-wired timer (not much different from the old Ranger sequencers) as a backup in case of failure. The Viking CCS had full redundancy; if a component failed, the computer would lose no functionality.

Another neat solution in the design was to "cross-strap" (cross-connect) most components. If a processor failed, the backup could come on-line and still use components (such as output unit, buffer, and power supply) that the failed processor had used—provided *they* hadn't pooped out. If they had, there were backups. The second computer only came into use during critical mission phases like Mars orbit insertion and midcourse maneuvers. The rest of the time it was unpowered, to save precious electricity.

Design for a Martian Computer

The actual design of the processor in the CCS was simple for reliability's sake. It used 18-bit instruction words made up of six instruction bits and 12 address bits. The machine thus had 64 instructions, and 4K of directly addressable memory. It ran at a "blazing" average of 11,000 instructions per second. One legacy from the Sequencer was the timekeeping. Every hour, second, and 10 milliseconds, an interrupt would pulse—activating routines that were time-dependent. The computer

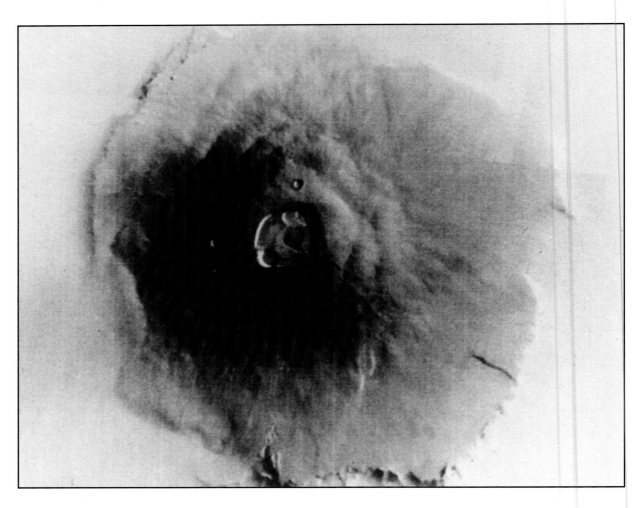

The crater of volcano Nix Olympica on Mars.

Image Processing

On any deep-space mission, the craft is stuffed with science experiments and sensors. These return varied information about magnetic fields, temperatures, sub-atomic particle streams, and other data on the environments of planets, moons, and space itself. Most planetary scientists and astronomers are more interested in this data than in the images that capture the popular imagination. The Pioneer series of probes did not carry high-quality visual imaging equipment, and it was nearly an after- thought on the Ranger and Surveyor lunar explorers. But when the general public remembers the Mariners, Viking, Voyager, and soon *Galileo*, they will remember the pictures: striking views of other planets.

In the beginning, the image-processing problem revolved around changing analog signals into digital form. The early space probes used television cameras with analog video. JPL developed a Video Film Converter to turn

continues

High-fidelity Voyager image of Saturn.

continued

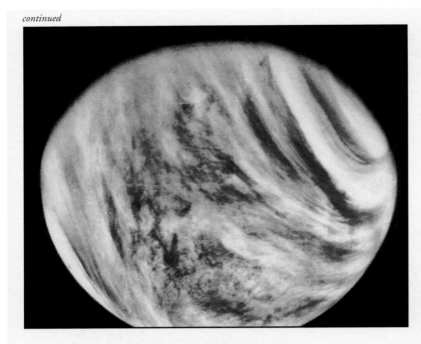

Late Mariner program image of Venus. Note the improvement over the image of the moon made by a Ranger probe.

Mariner 1964 visual data could only be transmitted at eight and one-third bits per second. It took over 11 hours to get one picture transmitted back to Earth. When Mariner Mars 1964 made its swing by the planet, it took only 22 of these images.

Once the data is received, computers enhance it. When you look at an image taken from a spacecraft, keep in mind that it has been "processed" by computers in subtle ways. If done to emphasize certain key elements of the visual data, image enhancement may actually distort other elements. Therefore a computer-processed image is less likely to reflect what you would see with your own eyes.

these signals into streams of bits. Later, digital cameras came into use. The basic unit of an image is the "pixel," or one dot. If you look closely at photos in some newspapers, you can see that they are made up of thousands of tiny dots, each with a difference in shading that makes the image stand out. Television sets also build pictures using dots, as do computer monitors. The *resolution*, or level of detail possible, is determined by the number of pixels: the more dots you use, the finer the detail.

Early Mariner images had only 200 rows of 200 pixels, about half the resolution of a typical TV camera of its day; the technology of interplanetary communication was in its infancy.

For instance, "contrast stretching" brings planetary features into unnaturally sharp relief. Suppose the spacecraft transmitted each pixel of

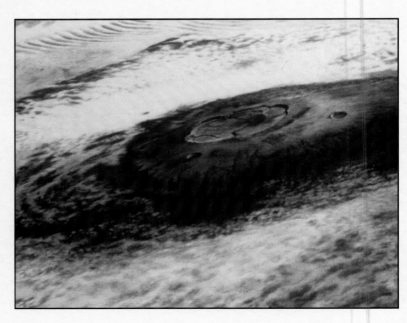

Viking orbiter image of Nix Olympica. Compare it to the image from five years earlier.

a picture as eight bits, and gave each a number from 0 to 255 (that is, ranging from all black to all white, with 255 shades in between). A software program would truncate the pixel values to six bits, shift the brighter greys toward the higher values, and adjust the darker greys toward the zero value. The result: a high-contrast image. Another program would "normalize" (make more even) the illumination of features in the picture, pixel by pixel. The program would take the average value of the 125 pixels nearest the one being processed, and subtract this average from the value of the pixel. Then the next pixel gets the same treatment.

By Mariner Mars 1971, the resolution had increased to 700 lines of 832 pixels—much better than the average TV of the time. Fortunately, the data transmission rate also went up, to 16,200 baud. Therefore a Mariner image could be transmitted in under six minutes; the ground controllers could watch the picture being "painted" in something closer to real time. All the image data could be kept for later processing as well, and now there was a lot more of it.

Mariner X pioneered color imaging, and the computing requirements increased even more. Then came the 1000-by-1000-pixel video camera, needing even more computing. Though the JPL spacecraft and their engineers led the way, image-processing technology has burgeoned into an industry in the decades since Mariner. Landsat Earth-imaging and weather satellites are now by far the most prolific sources of pictorial data. Commercial image processing is giving feedback to the space variety.

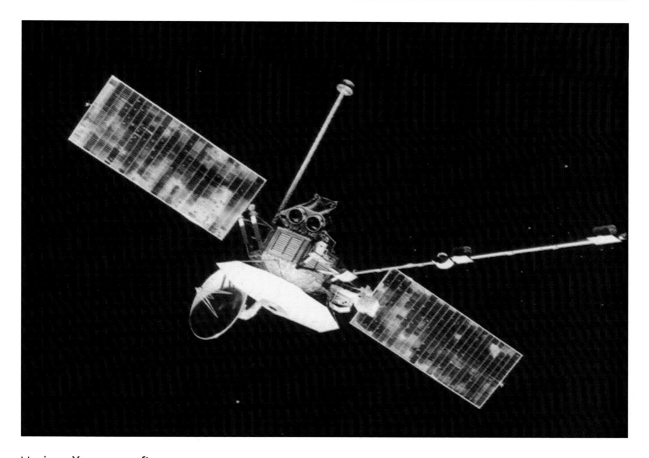

Mariner X spacecraft.

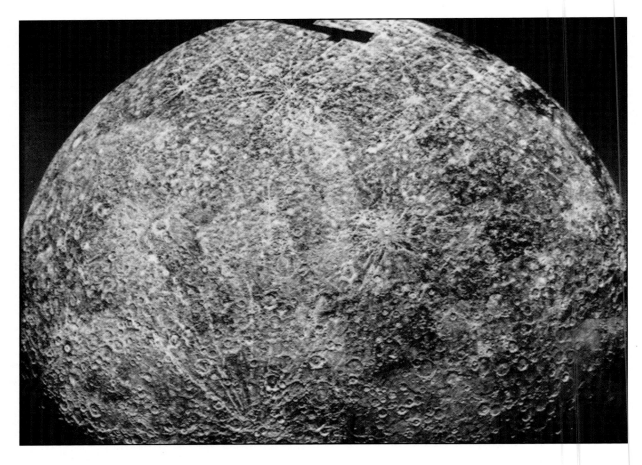

Mariner X images of Mercury, made into a mosaic.

also had a separate processor to handle hardware interrupts—32 levels of them. The software ran much like that on Apollo and the Shuttle—but instead of a (relatively large) software operating system, the Viking machine used hardware to handle the priorities and interrupts.

The memory was built of plated wire, which we met briefly as the memory of choice in the first version of the controllers developed for the Shuttle main engine (see Chapter 8). The programmers divided Viking's 4K plated-wire memory down the middle: a 2K read-only area would hold software, and a 2K read/write area would hold data.

Software Gains in Importance

Writing the software for Viking was both a new experience for JPL, and also vaguely familiar. After nearly a decade of struggling with alarm-clock timing code in the Sequencer, the software engineers had a chance to write real instructions with real processing.

The JPL team, led by Wayne Kohl, made many direct contributions to the design of both the hardware and the software. There wasn't as clear a distinction between the two (as on more recent projects, or with off-the-shelf hardware). The JPL programmers engineered the software extremely well (some would say because they were engineers by training, accustomed to building tangible things that worked). The Viking computer system would be a custom device from Day One.

The software and hardware groups had similar organizational structures and practices. Both closely followed a development plan that called for detailed documentation in specification, design, and implementation. They tested each routine before integrating it with the rest, but (as always with a real-time system), it was difficult to prove that all the parts would work together.

The software differed from that of systems like those in Apollo and the Shuttle. All input, whether interrupts or data, came into the same chunk of hardware, where it was sorted out. Each of the 32 interrupt locations contained an address that told the computer where to go and what to do if a particular interrupt occurred.

Once given an interrupt or data, the software would activate either a *command decoder* or the *event generator*. Both these blocks of software connected to each other, to the telemetry processor, and to the output blocks. The command decoder processed commands from Earth. The event generator contained the functions that had implemented the software in the old Sequencer; it made the spacecraft do things.

Basically, the computer had only two forms of output: telemetry to Earth, or commands to the flight and science systems. The computer and its software worked exceptionally well on a mission that took several years—the longest continuous computing run yet in space.

Software First: Viking's Lander Computer

Martin Marietta took a unique (and actually quite sensible) approach to developing the Lander's command computer. Unfortunately, their process has rarely been imitated since: they started with the software (and what they wanted it to do), and let the hardware catch up.

Over and over again, spacecraft computer systems and their memories had exceeded original projections, resulting in compromises. To achieve their mission objectives, Gemini, Apollo, Shuttle, Mariner X, and *Galileo* would all require additional memory or revamped software. Also, the capabilities of various processors had turned out to be too limited in some cases, forcing the abandonment of certain functions. This situation resulted from a "hardware-first" mentality that still dominates computer buying, even among present PC users.

Maybe Only on Saturday Night . . .

Life on Mars? The answer is "yes" and "no." I attended a colloquium celebrating the tenth anniversary of the landings, and heard two speakers using essentially the same data to prove that there absolutely *was*—and absolutely *was not*—life on Mars. The debate raged over chemical compounds that could be embedded in organic and non-organic matter.

Martin Marietta wanted to avoid being in a situation where the software was constrained by hardware chosen too soon. They got the contract in 1970, with launch originally set for 1973.

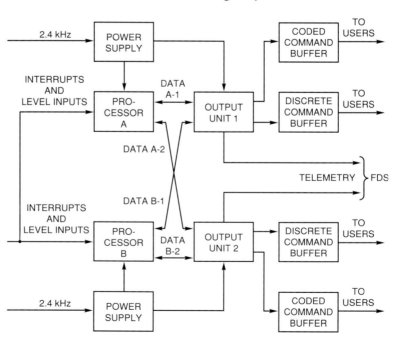

Block diagram of the Viking Command Computer Subsystem, showing the dual strings of input/output devices, power supplies, and processors.

Trenches dug by the Viking Lander soil scoop are visible in the lower right. The soil samples taken were processed by an on-board science lab to see if there were organic compounds present.

Then lean budgets delayed the launch a couple of years. Martin engineers knew that hardware would have improved considerably in five years, so rushing that decision was not necessary, or even smart. Their idea was to write the mission's requirements first, and analyze them to see what sorts of computer instructions would be needed to implement them.

Once they had their requirements, they bought a couple of Standard Computer Corporation IC-7000 processors. These special computers had a split CPU, and could be microprogrammed to emulate other machines.

The Viking team wrote microcode for one half of the IC-7000's split CPU: it would execute the Lander's proposed instruction set. The other CPU ran a simulation of the other Lander systems. In effect, half the computer would run the space-craft's programs, the other half would "pretend" to be the spacecraft. During software development, the IC-7000 could even be connected to the IBM 360/75 that JPL used for mission control. The Lander software could be tested as though it were in the actual spacecraft, communicating with controllers on Earth. When the Martin Marietta engineers found a deficiency that could be fixed by adding an instruction or two, they could easily do so.

The Martin Viking team was very disciplined. They wanted to hold memory growth to 18K at any one time. Some programs would be added after landing to replace the descent-control software (which would then be useless), but the plated-wire memory held 18K, period. Any time the software exceeded that size, it was cut back.

By 1974, Martin had a well-engineered, efficient, and clean software load. It was time to buy the computer. Unfortunately, other organizations contributing to the Lander did not have as much self-control. In the process of adding more science capability, the project engineer had let the weight of the experiments increase. He may have thought advances in computer hardware would provide a unit that weighed less than the original allocation, compensating for the added weight. That was not to be.

The computer that flew on Viking was actually the third best on the "wish list." It was short on a few instructions that the designers thought they needed, so the code had to be changed a little.

Nevertheless, it was the first use of an off-the-shelf processor in the deep-space-probe business.

The Viking team chose two Honeywell HDC 402 processors as the computer system for the Lander. These 24-bit systems averaged 230,000 instructions per second—pretty speedy compared to their predecessors. The computer ran on a 20-millisecond real-time "frame" length (an approach much like the time-slicing that the Shuttle's backup computer would later use). Its only drawback: software changes had to be hand-coded in octal (base-eight numbers, as you may recall from Chapter 5), making for some potentially ugly code.

Detail of the soil sampler in the Martian sunlight.

The computers ran the Viking Mission landers, now established as "Mars Surface Stations," for years. They could have continued, but had to be shut down; NASA was running out of money for the project, and wanted to move on. Someday, when humans explore Mars, a scene from the Apollo program (where astronauts visited a Surveyor lunar lander) may be repeated. If so, the Viking Lander will have waited a lot longer for its visitors than the Surveyor, and be a little more run-down because Mars has an atmosphere. But if the right interface could be found, the software would still be there in the plated wire. It would probably still work.

Even though the Viking dual-computer system was a big advance over the programmable sequencers that preceded it, NASA could not build and deploy from the Shuttle a spacecraft as complicated as *Galileo* without more experience. The agency would need to construct an entire flying computer network. The long-running Voyager program presented the opportunity.

Hardware First, Shoehorn Second

Many people ask the question, "What computer should I buy?" instead of the better question, "What am I going to *do* with my computer?" If they would find the software that fits their needs, and then buy a computer that runs that software, they would be a lot happier with the results.

The Viking Lander.

In the aerospace business, software is persistently acquired last. Some say this is because most big aerospace projects are run by managers who came up through the ranks doing "hard" engineering with tangible elements like electrical or mechanical components. The potential impact of software in modern integrated systems (in both cost and schedule) goes unrecognized—in part because computer-systems people are often the last consulted.

In Gemini, the designers of the craft had set the location and available space for the system (and thus the packaging constraints) before the computer and its software were procured. The Shuttle program would later decide definitely on a computer before the software requirements were even close.

There is one sure thing about requirements: *they will change*. Whenever computers and software are used in a new field, such as aircraft and spacecraft, applications rapidly increase. Everyone involved finds new things for the computers to do. Some hardware limitations can be overcome by clever software, but if the software is worked out first (and well) fewer such fixes are needed.

Martian landscape from a Viking Lander.

The Grand Tour of the Solar System

10

In the 1970s, an opportunity came up to do the astronomical equivalent of a hat trick, unassisted triple play, or back-to-back holes-in-one. A spacecraft on one mission, with clever navigational help, could fly by the four gas giants of the Solar System: Jupiter, Saturn, Uranus, and Neptune. It was a way to accomplish great exploratory feats—at the cost of a single launching, and a little over a decade of flight support. Individual missions to each of those planets would have taken generations at the budget levels of that time.

NASA began planning for the mission during the 1960s, and looked into various key technologies that would be needed to make it a success. One of these would be, obviously, reliable computers. Hardware failed frequently in the 1960s. Redundant flight control systems like those of the Shuttle were still in the future (they would have been too expensive and energy-hungry for an unmanned deep-space mission anyway). NASA searched for ways to achieve greater computer reliability without breaking the spacecraft's energy bank.

Composite Voyager images of Jupiter and its four largest moons, the primary objectives of the Galileo mission.

To the Stars with STAR

The now-famous reliability expert Algirdas Avizienis, a professor at UCLA, received a series of contracts to study the problem of computer reliability through the entire decade of the sixties. His most successful prototype was the STAR, or self-testing and repair computer.

The basic measure of reliability is Mean Time Between Failures, or *MTBF*. A component with a MTBF of 100 hours could be expected to run at least that long, but there was always the chance it would be one of those that failed in less than the average amount of time.

Avizienis understood that in most hardware, there are roughly as many below-average parts as above-average parts. He designed the STAR to have three copies of each component. That way there was a better-than-average chance of having a better-than-average component among the three. The STAR had three input processors, three logic units, three read-only memories (made of core rope from the same folks that did the Apollo memories), and so on. Every component was connected to every other through the Testing And Repair Processor (TARP). There were *five* copies of the very important TARP.

At first look, the design of the STAR (with all its extra components) seems to go against the requirement for conserving power. In use, however, only *one* version of each component was powered up at any one time. Three of the five TARPs had power, so that they could vote on data and instructions flowing through the system.

Most often, failures would be one of two types: permanent or transient. The TARP looked for both. Every word sent through the TARP was multiplied by 15 at the sending end and divided by 15 at the receiving end. If there was anything other than a zero remainder, something was wrong with the transmission.

The TARP maintained a *fallback point* ten instructions behind the current execution point. If there was an error in transmission, the software jumped back to the fallback point, and executed the instructions again. If the error was transient, then the software would continue its work. If the error was repeated, there was a good chance the transmitting component had failed; it would be turned off, and a "new" one brought on line.

The TARP was a little harder on itself than it was on the components under its supervision. If a vote was not unanimous, the dissenting TARP was shut down, and another brought on line. After the second TARP failure, a previously-failed TARP would come back on line, since chances were good the original error had been transient. If the error was permanent, the TARP that had it would simply be outvoted. Four of the five TARPs would have to fail to put STAR out of business.

STAR never got a chance to be packaged to fly in a spacecraft. It turned out to be too big. But some people also doubted whether a system could actually *be* more reliable if it were always turning on and shutting down components. It did offer gains in the total reliability of its components, but it also depended more heavily on switches—and they could become single points of failure unless they too were made triple-redundant. Eventually the search for complete reliability foundered (as it often does); concurrent increases in size, power, and weight—those three worries of every spacecraft designer—were nearly inevitable.

Even though STAR never proved itself, the concept of keeping redundant components dormant *did* make an impression on JPL and NASA. Both spacecraft of the Viking probes had redundancy. There was no way a Grand Tour mission could avoid it.

A Computer Network for Deep Space

Just about the time STAR began running in its "breadboard" (experimental) version, JPL began designing the Thermoelectric Outer Planet Spacecraft, or TOPS.

TOPS differed from the Mariners in the source of its power. The Mariners used solar cells to

The ringed planet Saturn, second of the outer planets beyond the asteroid belt.

generate electricity. Since they did not venture past the inner solar system, these did not have to be very large. But a spacecraft en route to the outer planets had to pass a kind of threshold. This was the asteroid belt—the Solar System's ring of planetary debris beyond the orbit of Mars. Once past the asteroid belt, solar radiation would drop off so much that solar panels would have to be enormous to produce enough power. So a nuclear power source was proposed instead.

The new power plant for the unmanned spacecraft would carry a small amount of plutonium. The radioactive plutonium would excite the heat of thermocouples, causing them to generate electricity. Radioactive elements decayed as they gave off radiation, a process called the element's *half-life*—and plutonium had a long one. As a power source, it could make very long missions possible. That was the good news. The bad news was that the radiation could "cook" the spacecraft's circuitry; the power source would have to be mounted on a

boom well away from the electronics. This would add some weight and "unbalance" the probe.

As funding for NASA dwindled in the post-Apollo days of the early 1970s, TOPS was cancelled—and the Grand Tour mission with it. JPL came back with a proposal called Mariner Jupiter-Saturn 1977, a sort of less-than-Grand Tour of the two nearer gas-giant planets. The JPL engineers kept a hopeful thought in the back of their minds: on a certain trajectory, a well-built spacecraft could still make the Tour. With a healthy probe, and enough energy from gravity assists to make the Uranus and Neptune encounters, it would be harder to justify shutting down the mission. As the project progressed, it got a new name: Voyager.

In many ways, Voyager was a lot like previous JPL probes. Like the Mariners and Viking, it was designed to have a command computer. Some aspects of TOPS became part of Voyager because there was little choice—in particular, the nuclear power generators. This created a need for yet another computer—to control the attitude of a spacecraft that would have to mount equipment on booms.

"Hold That Pose"
All JPL spacecraft flew in *three-axis inertial mode*. This meant that the spacecraft was always oriented in a particular way. It made it easier to keep a communications lock on Earth and to point the cameras and other sensors.

Previous spacecraft designs had featured booms of sorts, but all the really heavy components had been mounted on the "bus" (the main structure); concentrating the mass allowed for more stable flight. Voyager too would need booms, but they posed problems. The thermoelectric generators would have to be mounted on one. Heavy and far from the spacecraft's center of gravity, they would tend to throw its mass out of balance. JPL had also decided to mount the scan platform—with its cameras—on another boom. That way the cameras could be traversed (turned as needed) without moving the core part of the spacecraft, but the addition did not improve maneuverability. The two booms flexed any time the attitude changed; engine firings and motion of the cameras would

make them vibrate in powered flight and in encounters. Pointing and steadying the parts of the craft would be more difficult.

Voyager's needs were clearly more complex than those of its predecessors. Mariner and Viking had accomplished their attitude control with hard-wired logic, but this mission presented new challenges. Voyager had to handle the difficulties presented by the booms, point the complex new scan platform, and help provide guidance to the "kick stage" rocket that would complete the launch phase and send the spacecraft on its way. To meet these needs, JPL added computers to the attitude-control system.

As Voyager's design took shape, the engineers realized the old hard-wired data formatters would be overwhelmed with the volume of information Voyager would gather. They decided to add computers for the data system as well. Finally, three sets of computers (two redundant computers per set) would fly in an on-board network on each Voyager spacecraft. Until *Galileo*, this would be the largest network of computers in an unmanned spacecraft. Both Voyager I and II took off in 1977; by 1994, their computers would have over a decade and a half of continuous use behind them.

Control Computers on Voyager

The Central Computer and Sequencer (CCS) on Voyager had a remarkable resemblance to the one on Viking. No surprise here; it was a deliberate attempt to standardize. A flight computer system much like the CCS had been standard across several Mariner missions. It saved money and time.

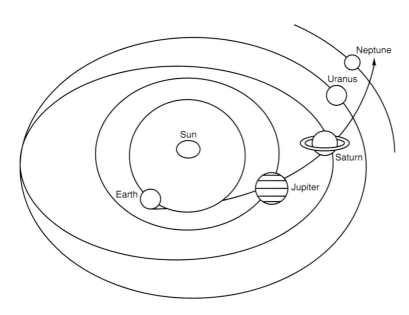

The Grand Tour of Voyager II.

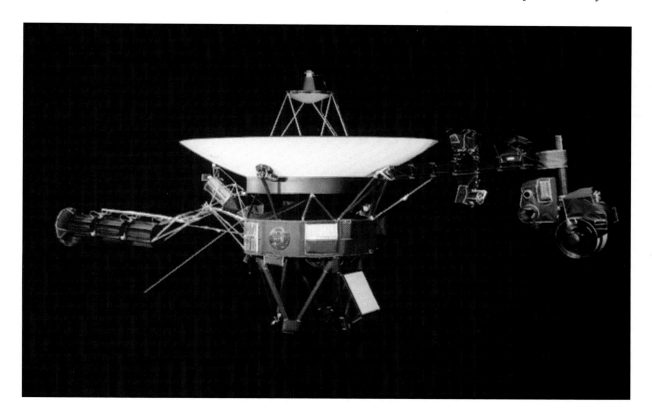

The Voyager spacecraft.

This version of the CCS, however, had some major changes. It had added interfaces to the Attitude, Articulation, and Control System of the spacecraft, as well as to the processors of the Flight Data System. Any commands for these systems would come in from Earth, and be routed through the CCS. Even with the new functions, the memory stayed the same size as on Viking: 4000 words. JPL's practice of overwriting "old" software in flight had become so standard that there was no real reason for larger memory (just as your home computer does not need more memory if programs can be run one at a time).

Even so, the software engineers were always looking for ways to increase the capability of the computer system. A typical example of their quest would come well into the mission: one of the Flight Data System memories on Voyager II had a failed bit in a readout register, permanently indicating a "1." CCS programmers tried to use the "good" parts of the memory for auxiliary storage.

Maintaining a Good Attitude

The Viking Lander mission had proved a computer could be used for attitude control; a Honeywell machine, programmed by Martin Marietta, had assisted its Mars descent. For Voyager, the computer's role and capabilities would expand again.

In preparation for Voyager, JPL did a feasibility study of computer attitude control called HYPACE. The "HY" stood for "hybrid," because the design used a combination of digital and analog devices to maintain pointing. When the time came for Voyager, however, budgets ruled supreme; a new computer like the one recommended by the HYPACE study was out. Instead, the command-computer folks convinced the attitude-control specialists that the CCS machine would work for *both* jobs. General Electric, which manufactured the CCS, would build the new computers; Martin Marietta would finish off the rest of the system. It was called HYPACE in all the Voyager documents, even though it was not the exact device from the test program.

Launch of a Titan III booster with the big shroud used to enclose the Viking and Voyager spacecraft during ascent through the atmosphere.

Voyager II Survives a Bad Attitude Problem

Shortly after separating from the last rocket stage during the launch phase, Voyager II lost its planned attitude—just as the booms began to deploy. The plumbing for the thrusters was in the process of being reconfigured, so the attitude-control system could not immediately correct the problem. Since this was a critical mission phase, both computers were up and running; both dutifully issued the reconfiguration command—over and over. A reconfiguration would be completed, only to be done all over again, delaying the response to the attitude problem even more. Finally, the "prime" computer gave up and skipped a heartbeat. The backup went on line as a single processor. It executed the standard orientation routine, and gave Voyager a more righteous attitude.

Adapting to New Duties

The CCS computer needed some changes if it were to be used for Voyager's attitude control. The CCS ran much too slowly to control attitude during the kick-stage burn—it needed to be roughly three times as fast. Other tasks (such as controlling the cameras' stepper motors on the scan platform) had to be done quickly too—more quickly than the CCS could manage.

Other things were less time-critical. The software for the system ran in four time groups. Thruster firings and scan motors were controlled in a 10-millisecond frame. The calculations for attitude control and use of the thrusters were in a 20-millisecond frame. Control of scanning was in a 60-millisecond frame. Finally, the command interpreter ran every 240 milliseconds.

Every eight cycles through the 240-millisecond frame the CCS sent a "heartbeat" signal to the command computer—a set of codes to tell the command computer what was happening in the attitude-control system. (The signal was carried on a six-bit direct connection, similar to the one that would link the computers on the Shuttle.) Between heartbeats, Voyager's computer conducted self-tests. Code 66 (called "The Omen" for reasons film buffs will recognize) warned the CCS to *save all parameters, the computers were coming down.* Other self-test failures caused the heartbeat generator to be skipped. If the CCS detected no heartbeat for 10 seconds, it assumed the current attitude-control computer must be dead, and sent commands to activate the backup. This use of the CCS kept redundancy management outside the computer system, protected by the reliability of plated wire.

As the years of flight went by, the attitude-control computer system would be given the task of extending the aging spacecraft's useful life. By the

The Tale of the Lonely Voyager

The software architecture for Voyager's CCS was essentially the same as Viking's CCS. The Event Generator module carried a routine intended to be a safeguard against losing contact with Earth. This was called CMDLOS, short for "command loss." (Note the six letters. Everyone associated with the software apparently had FORTRAN-on-the-brain, since older versions of FORTRAN only allowed six-character names.) The CMDLOS routine had a parameter that could set one of the familiar alarm clocks.

The Voyagers would be in cruise mode far longer than in any other mission phase; JPL intended to communicate with them only at intervals, to receive science data and send up new commands. Since radio contact would not be continuous, the spacecraft would know it had lost contact with Earth only if it missed an expected command. The parameter in CMDLOS corresponded to the time of the next command—say, seven days. If the alarm went off, the spacecraft would go through an emergency procedure to re-establish contact. To keep this feature current, resetting the CMDLOS alarm parameter would be one of the first commands uplinked during each communication from Earth. The emergency procedure was similar to a loop. First, the computer would try to figure out whether the probe's radio antenna was still pointing at Earth.

This information was vital. Over the long distances involved, even a very powerful transmitter could only deliver a very weak signal at the

continues

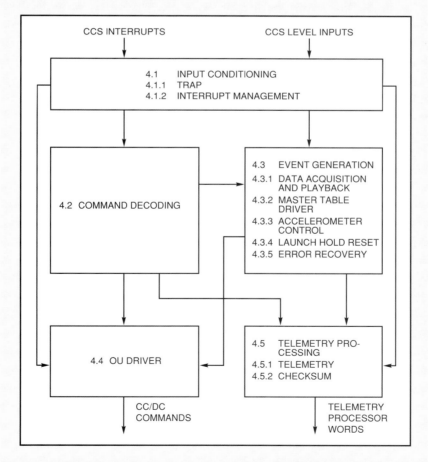

Structure chart of the Voyager Command Computer Subsystem software.

continued

receiving end. Voyager carried a directional antenna (like a backyard satellite-receiver dish) that could be adjusted to get the best signal—and it had to be facing Earth to get anything.

If the CCS determined the spacecraft was off its correct attitude, it commanded the Attitude and Articulation computer to re-orient the spacecraft. In response, sensors would align with the Sun and a bright star like Canopus, determine the angle to use, and point the antenna at Earth.

Once this realignment was complete, the spacecraft was to "assume" that the Mission Control computers would be sending a stream of commands, frantically trying to get the spacecraft's attention. If no commands were detected, the spacecraft's next assumption would be that its radio receiver was dead. (The logic was that if the ground-based transmitter were broken, the controllers would rush to their friendly local radio parts store and get things to fix it. Therefore, any broken radios must be on the spacecraft.) The CCS would then shut down radio number one and switch

on radio number two, following the same policy of redundancy and dormancy that ensured the computer system's reliability.

Before the first Jupiter encounter, Voyager II's CCS had a CMDLOS alarm. After reacquiring Earth and switching radios, it sent telemetry indicating what it thought had happened. There was no response from Earth. In such a case, the CCS was to repeat the entire re-orienting and radio-switching process. Eventually the Deep Space Network pointed antennas at Voyager II and heard its telemetry barrage. A command was immediately uplinked to reset CMDLOS. Unfortunately, all the switching back and forth had caused the number one radio to fail. Since Voyager I's trajectory would not be usable for the Grand Tour, Voyager II was the spacecraft designated to do the Uranus and Neptune encounters. That mission was in jeopardy; there was only one radio, and over a decade of flying remained.

What had happened? The controllers had simply forgotten to send a command in time to reset the CMDLOS parameter. It was a human error.

time the Saturn encounter ended, the actuators on Voyager II's scan platform were just about shot. Special code had to be used to point the cameras during the Uranus and Neptune encounters. This is no mean feat. Something like a two-word fragment of free memory had been available in the 4K memories at liftoff. Any changes in code came hard.

The Homework Assignment That Flew to the Planets

The design of Voyager posed yet another challenge. The imaging work and science experiments planned for the mission meant the data flow

through the computer would be much higher than on any previous spacecraft. There was some brief consideration of using the CCS to provide some formatting and data-handling, but it did not take long to realize that the volume of data would be too much for it. A custom-built computer would be needed. JPL's Jack Wooddell got the job of designing one.

Conveniently, Wooddell was working on a graduate degree at USC that included courses on computer engineering. His assignment to design a computer for the Flight Data System (FDS) coincided with a course assignment to design a computer. Wooddell did his paper on Voyager's FDS computer. His study laid out the functions required of the new machine, and also the likely restrictions on its reliability. He then divided the computer's functions: some would work best as hardware, and some would work best as software. Once he had decided which functions would be which, he was ready to design an appropriate set of instructions for the computer. Enter Dick Rice, a master programmer.

Rice wrote programs for the breadboard version of Voyager's Flight Data System. He discovered that to obtain the needed data rate, he would have to double the processing speed. One way to do that was to add direct memory access to the machine. Wooddell modified the instruction set, adding a direct memory access to those instructions that did not have one as part of their execution cycle. The result: now the entire instruction set consisted of instructions with execution cycles of equal length. That made execution a lot easier to predict.

The computer was also speeded up by using semi-conductors for both the processor and the memory. This was a *major* change; semiconductors needed continuous power, and spacecraft were power-stingy. Wooddell suggested CMOS-technology chips, which were nearly new at the time. They were low-powered and tolerant of varying voltages, characteristics attractive to designers of spacecraft computers.

The problem with CMOS memory was that it would be "volatile"—it would need to be continuously powered to retain the data and programs. Fortunately, a backup power supply turned out to be easy to obtain.

The spacecraft's power would be provided in the form of *alternating current* (just like your household electricity). In addition to that power supply, a special cable ran to the thermoelectric generator that produced *direct current* from the thermocouples; it would keep the memory "alive" in case of

The first imaging of Uranus, taken from 46 million miles out.

Neptune, with its big moon Triton in the foreground, as imaged by Voyager II.

Quality Pictures, Fewer Bits

Once Voyager had finished the Jupiter and Saturn flybys, the FDS programmers (and their image-processing counterparts) needed to solve a tough problem: how to maintain the breathtaking image quality during the Uranus and Neptune encounters? The problem was that the farther out Voyager flew, the more the transmission rate had to drop. At those distances, its radio did not have the power to transmit cleanly at the same rate.

The spacecraft used tape recorders to store the imaging data, and then transmitted it through the Flight Data System. The FDS, in turn, formatted the data and hosed it toward Earth. In the meantime, the tapes were being refilled with new bits. If the transmission's data rate slowed, the tape recorders filled more quickly with imaging data that was not being sent, and imaging had to stop. There were two alternatives: increase the data rate somehow, or take fewer pictures.

Recall that the imaging data is transmitted in the form of a value for each pixel. In Mariner, Viking, and early Voyager, the way to get high-resolution images was to send the value of each pixel in its entirety. Studying the problem, the engineers noticed that the difference in value between two adjacent pixels was not too great. In a typical image, there were fewer sharp lines of light and color demarcation than there were spaces of similar color and illumination. They hit on the idea of having Voyager transmit the full value of the image's upper-leftmost pixel, and after that, send only the difference in value between it and the next pixel in order. The difference would usually be so small that it could be sent using less than one-third the bits that would have been needed to represent the pixel's true value.

The calculation of these differences was a job for the FDS. JPL put the second processor to use doing this work. The resulting data compression made deep-space images like those in the figure possible.

an electrical outage. If something happened to the spacecraft that was bad enough to knock out the generators, the mission would be over right then; the data computer would not be needed anyway. This was the first time semiconductor memories flew in space. (The fact that Voyager's are still working today—after over a decade and a half of flight—no doubt reassures the designers who used semiconductors in the Shuttle's computers.)

Eventually the 16-bit FDS computer reached a speed of 80,000 instructions per second. Because using CMOS technology had saved size, power, and weight, Wooddell was able to expand the memory size to 8K. Ed Blizzard joined Dick Rice in 1975, and they began to work on the flight version of the FDS code.

A Real Programmer . . .

Some years ago, the programming underground press circulated an essay titled "Real Programmers Don't Write Pascal." It was an ode to the days when coders were shamans. At one point, it praised a "programmer at JPL that had the entire Voyager flight code memorized." Near as anyone can tell, the anonymous programmer is Dick Rice, who worked on the FDS for more than a decade. Of course, he only had the FDS flight code memorized, but that is enough for a legend.

Rice says the basis of the software's architecture was *functional*. There was no top-down or structured way of looking at the collection of functions. (We'll restrain ourselves from claiming they

were doing "object-oriented design," ten years before the term became popular.) The 8K of memory was divided along the usual lines: one half for programs, one half for data. To get more space for data, they finally decided to have one processor access the "data half" of both memories. The program halves would be kept in duplicate. This way the second processor could pick up if the first one failed, without having to rebuild all the code and data.

On to the Stars

After the Neptune encounter, Voyager I left the plane of the ecliptic (the plane in which most of the planets orbit). Even as you read this, this highly successful probe is searching for the edge of the solar wind—the point at which the actual physical pressure of the Sun's radiation becomes undetectable—the true boundary of the Solar System. Voyager II successfully finished the Grand Tour, and is out of the imaging business, also leaving the Solar System. Will one of them meet the real equivalent of Admiral Kirk sometime in the distant future? Not very likely, but they certainly would make a fine exhibit in some interplanetary computer museum: Earth's first distributed computer system in deep space.

After Voyager headed out on its endless flight, NASA knew enough about tying different types of computers together in long space missions to take on the task of designing and building *Galileo* (which is, also as you read this, on its troubled way to Jupiter).

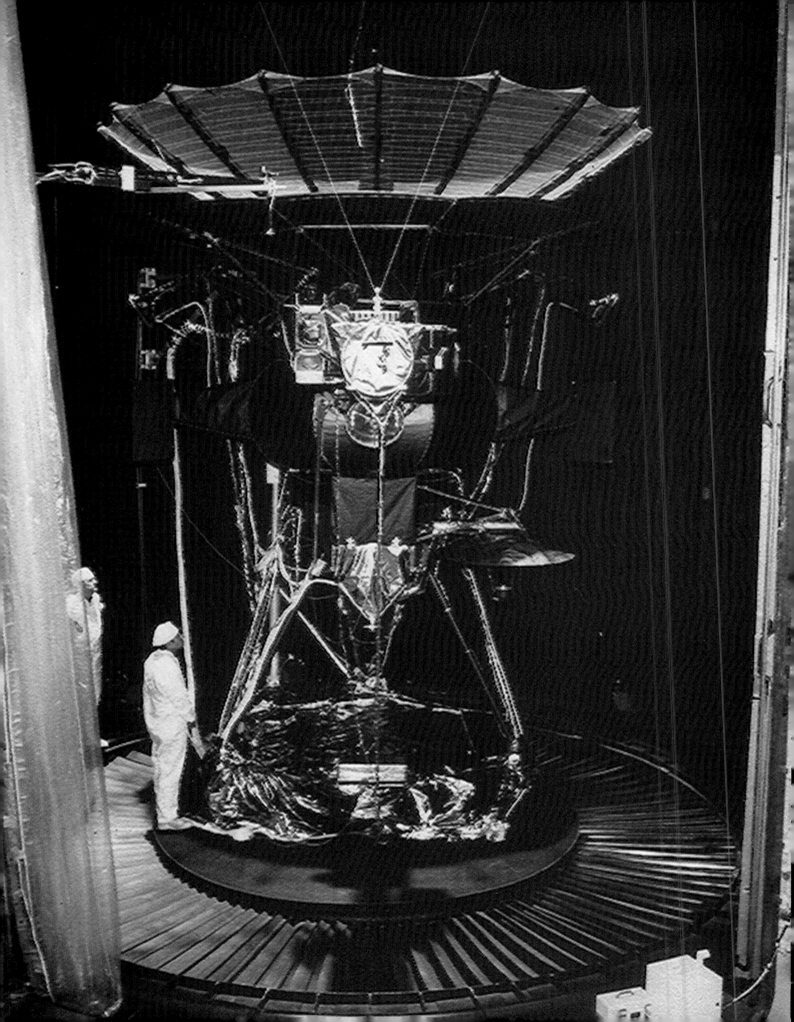

Galileo's Decade of Flight

11

The *Galileo* spacecraft launched by *Atlantis* to Jupiter
has had even more trials than its namesake. The astronomer and
physicist Galileo was persecuted because he thought the
"geocentric universe"—the idea that Earth was the center of the
universe—was wrong. The first to see the moons of Jupiter, he rightly
deduced that the Earth-moon-Sun system was similar. His work was
attacked by everyone from fellow scientists to the Church. It took
centuries to officially clear his name. Of course, it helped that he was
absolutely right.

Galileo and its design team suffered through many delays and
launcher changes, making it difficult to get software ready on time.
There was also an added pressure: the spacecraft was to be a one-
shot deal. Vikings and Voyagers had flown in pairs; if one failed, the
other could complete the mission. It had not been necessary to redo
all the ground-control work and rebuild the spacecraft. *Galileo*,
however, did not have the "luxury" of a backup spacecraft.

Customarily a second probe could be built for less than 15 percent
of the first one's total cost; the expensive engineering usually went
into figuring out how to build the first one. But NASA was too broke
to build a second *Galileo* (or to launch a single test model), in spite
of the great risk of failure.

The *Galileo* probe during pre-flight testing.

Galileo's great complexity—and the sheer number of things that could go wrong—made the lack of backup especially dangerous to the mission. The complexity had grown partly from a desire to please different groups of scientific investigators, and partly from a need to build in redundancy. *Galileo* would have to be ultra-reliable.

If you compare *Galileo* to Voyager, some things look familiar. There is the big high-gain antenna. There is the thermoelectric generator on its boom. Other things are different. There is the slender magnetometer boom—longer than Voyager's—and the blunt object hanging underneath the main spacecraft bus. Those two items emphasize the complexity of *Galileo.*

The blunt object is a probe to investigate the Jovian atmosphere. It is to be released five months before the remainder of the spacecraft goes into orbit around Jupiter. The plan is for the probe to descend into the swirling gases of the giant planet, and transmit data until it fails, crushed under gas pressure or zapped by radiation. The remainder of *Galileo* will relay its data to Earth as it enters orbit. After that, it will spend the next few years gathering data about the *Jovian system* (as astronomers call Jupiter and its satellites), and making close fly-bys of its many moons.

The upper part of the spacecraft (with the magnetometer boom) is designed to spin three times per minute. This allows the sensitive magnetometer to tell the difference between electrical fields in space and those in the probe's electronics. NASA's Pioneer probes (launched in the 1970s) were also "spin-stablized," so it is not the first time this has been done. The lower section is three-axis inertial, just like the Voyager (see Chapter 10). The two sections are designated "spun" and "de-spun."

Work on *Galileo* started in the late 1970s, with the intent to launch it from the Shuttle in 1982. Delays in the Shuttle program forced the date back into 1984, then 1985, and then May of 1986. Unfortunately, the destruction of the

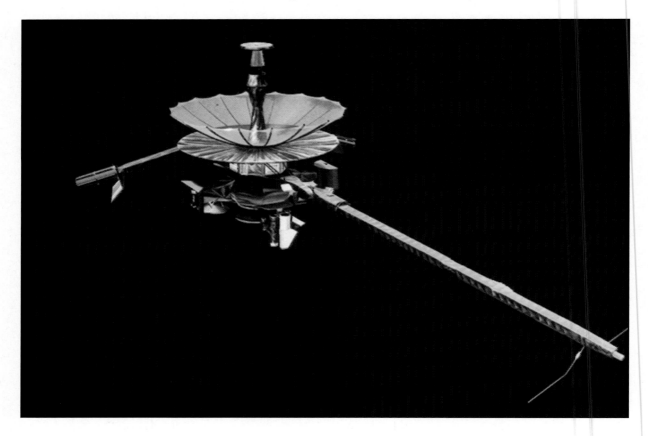

Galileo deployed.

Challenger orbiter suspended flight operations for a couple of years; the launch date pushed back to the fall of 1989.

The original intention was to have *Galileo* fly a direct path to Jupiter, using a Centaur liquid-propellant upper stage. The Centaur had been around since the 1960s. It was the stage below the kick stage on the Voyager launches. For the Shuttle version, it was to have 50 percent more fuel. It also had some modifications to adapt it to the cargo bay of the orbiter. In the post-Challenger safety orgy, the Shuttle-Centaur idea was cancelled—the vision of doing a return-to-launch-site abort with 20 tons of liquid hydrogen and oxygen in the cargo bay was not attractive. The Inertial Upper Stage (IUS) replaced the Centaur. Having the advantage of being solid-fueled (harder to ignite accidentally), the Inertial Upper Stage would fly frequently on the Shuttle.

Unfortunately for the cost of mission control—and for the patience of the scientists—the Inertial Upper Stage was so much weaker than the Centaur that *Galileo* would have to do gravity-assists at Venus, and (twice) at Earth. It would now take six years, instead of about two years, to make the transit to Jupiter. When *Galileo* arrives at Jupiter in late 1995, it will

An Atlas-Centaur launching in the 1970s. Still in service, the Centaur is the upper stage mounted on the top of the Atlas.

be 11 years after the orbital-insertion date planned originally.

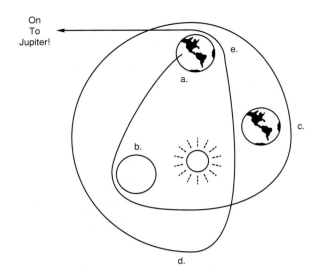

a. Launch, 10/89
b. Venus flyby, 2/90
c. Earth flyby #1, 12/90
d. Thruster burn, 12/91
e. Earth flyby #2, 12/92
f. Arrive Jupiter, late 95

The modified flight path, helping Galileo obtain multiple gravity-assists out of the inner Solar System.

The changed mission profile meant at least one obvious change in the design. The spacecraft had to spend half its flight in the *inner* solar system doing the gravity-assist maneuvers. The original design—intended for deep space—had not needed much insulation from heat and radiation. After all, the original mission had called for the spacecraft to leave the Sun even farther behind. The original design did not provide sufficient protection from solar energy for the main spacecraft bus. A sun shield was added to protect it.

Requirements for the Computing System

Galileo would have the distinction of being the most extensively computerized space probe in human history. The usual mission requirements applied: launch, maneuvers, cruise, and encounter. Therefore command computers would be needed. There was no doubt that attitude-control computers would also be needed; the spacecraft's balance would be affected by its

"spun" and "de-spun" sections, as well as by deployment of the atmospheric probe.

By the 1980s, principal investigators had realized the utility of microprocessors, and had included them on nearly every experiment *Galileo* was to carry. This required some increased capability on the part of the spacecraft data-processing system. In order to meet these needs, the Jet Propulsion Laboratory (JPL) was able to take advantage of some of its own basic research into computer architectures.

Unified Data System: Son of STAR

One of the young computer scientists working with Professor Avizienis on STAR (Chapter 7) was David Rennels, also of UCLA. In a research project sponsored by JPL, Rennels used the increasing capabilities of microprocessors to experiment with a new architecture.

**Technology Transfer:
Slow But Steady**
It's interesting how these concepts make their way from space applications into other real-time systems. The popular imagination assumes direct transfers of technology from space projects to industry; though not many actually occur, some sort of transfer is always going on. The Unified Data System "won" the competition to fly on *Galileo* over the Voyager system and the NASA Standard Spacecraft Computer (an IBM machine used in Earth-orbiting spacecraft such as Solar Max and the original design for the Hubble Space Telescope). That is clearly a direct infusion of research results. Some concepts of processor specialization and reconfiguration also showed up in military real-time designs (such as the RAH-66 Comanche helicopter) and the commercial Airbus A-320 transport.

Studies by the Air Force and others had shown that avionics systems worked better when their tasks were distributed functionally. The failure of a single processor in such a system would have a less drastic effect on the whole. Having parallel processors in the system took better advantage of the available computing power. Rennels experimented with various ways of increasing the system's flexibility.

He came up with a two-tiered computer network. Each High-Level Module of this system had its own processor and memory. The High-Level Modules provided the functions needed by the entire system, such as bus control and error detection. On the lower tier, the processors were called "terminal modules." They connected directly to spacecraft engineering systems, each controlling one functional area.

All the computers in this network accessed the memory directly to communicate with each other. The High-Level Modules "talked" only to the terminal modules, and to each other; they could not deal directly with a subsystem or a scientific experiment. This architecture came to be called the "Unified Data System." It continued some of the same features that had first enchanced the STAR computer's reliability. There was no central bus controller—each High-Level Module controlled a *separate* bus, so that data could still be transferred in case of multiple Module failures. Any of the High-Level Modules could transfer data from any memory, *to* any memory. After a failure, the remaining High-Level Modules could take over the functions of the broken processor.

The beauty of all this was the extreme specialization. There was a system-wide clock signal every 2.5 seconds, but specific tasks never had to share a computer. That meant the individual processors could be programmed without excessive worry over synchronization.

Alphabet Soup: HLCs, LLMs, BUMs and DBUMs

The JPL software engineers did some preliminary construction of expected *Galileo* programs, experimented with allocating functions to various computers, and tried out some new programming languages. They hoped to use HAL, the programming language used in the Shuttle's on-board software, for all *Galileo* systems. They also wanted to use a commercial microprocessor (instead of developing still more custom hardware).

The first decision they made combined the Voyager Command Computer Subsystem and Data Processing System into the Galileo Command and Data Subsystem (CDS). JPL had wanted to combine these functions as early as Voyager, but the command computers had been too weak. With *Galileo*, however, it would be possible to spread the software across many microprocessors. That meant it was possible to combine these functions.

The microprocessor turned out to be a compromise choice: the RCA 1802. RCA 1802s were eight-bit machines, like most early

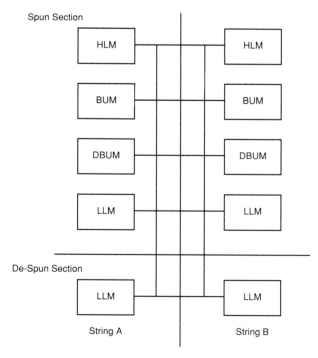

Schematic of the Galileo Command and Data Subsystem.

microcomputers. They executed an average of 100,000 instructions per second. This was slower than hoped, which made some aspects of the software development difficult. However, the RCA 1802s were CMOS devices, just like those in the Voyager Flight Data System; they needed only 30 milliwatts of power.

The original *Galileo* architecture used three of these processors as High-Level Modules (HLM), and four more as Low-Level Modules (LLM; similar to the terminal modules in the Unified Data System). There were three redundant buses, each controlled by a separate High-Level Module. The Low-Level Modules shared the usual data chores of an unmanned spacecraft: sequencing, telemetry, self-testing, and the like.

Dynamic Reallocation: Non-Stop Software

When programs are spread over many computers, the loss of one computer ought not to mean the end of processing of the software on it. An operating system can be designed to sense the failure of a piece of hardware, and boot up a new version of the software in an under-utilized computer on the network. Such systems often have at least one common memory area, where recent values of parameters are kept. That way, the newly-started replacement program can go to the memory and get information that enables it to pick up (more or less) where the failed computer left off.

JPL engineers eliminated one of the HLMs in the final Galileo Command and Data Subsystem. The bus it controlled was kept for use during testing, but would be disabled in flight. That left two buses: the central components of the rearranged architecture.

The UDS pioneered a new technique for dealing with a hardware failure: the *dynamic reallocation* of software resources. The CDS kept some aspects of

dynamic reallocation, but was now organized into two redundant strings of equipment on the buses. Each had a single HLM and two LLMs. The HLM and one LLM were in the spun section of the spacecraft, the other LLM in the de-spun section. The bus used a rotary transformer to bridge the connection between the two parts of the spacecraft.

The software had to account for resulting signal delays between the spun and de-spun sections. The HLMs had 32K memories, and the LLMs had 16K each. In addition to the memory that was part of each computer, there were Bulk Memory (BUM) and Data Bulk Memory (DBUM, pronounced "dee-bum") modules. The BUMs acted primarily as command buffers, the DBUMs as data buffers. In the spun section, there were dual 16K BUMs and 8K DBUMs. The result was a six-processor network with 176K of semiconductor memory.

The redundancy on *Galileo* was handled much like that on Voyager. Both strings would be active for critical mission phases. Only one would be up during cruise. Each could turn off the other line if it detected a failure. By handling the redundancy at the string level (instead of at the processor level), the entire system was simplified—as least it was now as simple as it was going to get.

Software for the CDS

A dozen software engineers worked for a decade to get the initial Galileo Command and Data Subsystem code working well. At the beginning, the engineers shared five terminals connected to an IBM 370/158 mainframe computer.

JPL failed in an attempt to make a HAL/S compiler suitable for generating machine code the 1802 processors could use. The Laboratory then decided to create a set of assembly-language macros (miniature programs), and document them as sort of an intermediate-level "programming language." This was an approach similar to using an *interpreted* programming language for the Apollo

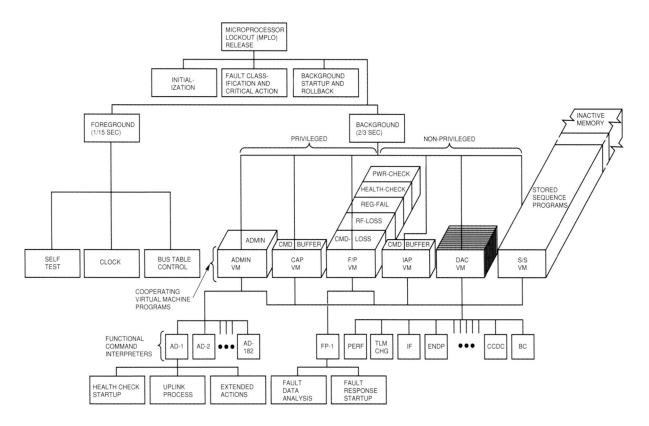

Command and Data Subsystem software structure.

spacecraft computer. The macros had familiar names like "IF," "ELSE," and "ASSIGN"—and others not so familiar. Each time one of the names appeared in the code, a set of machine-level instructions would be executed just as if it were the output of a compiler.

In the HLMs, there are two groups of software modules; each ran at a different frequency. The "foreground" processes included the clock, the self-testing software, and the bus controllers. They ran 15 times a second. Every two-thirds of a second, a longer list of "background" processes would be executed. These background processes were divided into six "virtual machines." (It seems that for every "real" thing in computing, there is a "virtual" thing. In this case, the virtual machines enabled spacecraft tasks—such as command processing, fault tolerance, and stored sequences—to be kept together.)

Making Big Computers Out of Small Computers

In the beginning of the microprocessor era (the early 1970s), the first "computers on a chip" used words four bits in length. Many computer designers combined two such chips to make 8-bit machines, and four to make 16-bit machines. Special registers that held 16 bits were often installed on the processor boards to act as central-processing-unit (CPU) registers. *Accumulators*, which held the current total of the calculations, and *address registers*, which contained the next memory location to be executed, are examples of *CPU registers*.

The six virtual machines were further separated into "privileged" and "non-privileged" machines. If a non-privileged machine did not get to run

Galileo in final assembly.

processing machine containing some old favorite routines (such as CMD-LOSS). If something in the fault-detection software sent up a red flag, contingency command sequences would run to handle it.

The plan was to avoid changing the software in the privileged machines after launch. The regular weekly software update would be in the non-privileged machines. The two virtual machines would run immediate-action and delayed-action command sequences; a third would maintain stored sequences for later distribution to the computers. This was the "dangerous" side of the software; since it would be constantly changing as the mission progressed, there was more room for error to creep in.

The Balancing Act

Galileo had one other main spacecraft computer complex: the Attitude and Articulation Control Computer System (ATAC). The main requirements for the computers in it were speed, speed, and speed.

The attitude-control problem was tougher (and the imaging rate faster) than on any previous unmanned spacecraft, creating the need for greater processing speed. After looking at many alternatives, JPL settled on the Itek (now Litton

before the end of the two-thirds-second cycle, its operation could be canceled until the next cycle. The privileged machines could never be canceled. This way if the processor workload was extreme, critical functions would still get done.

As you can see in the figure, the three privileged machines ran various things: the administrative (executive) software, a command buffer containing contingency plans, and a fault-

Industries) 2900 series 4-bit microprocessor. Some versions of it had flown in naval aircraft, proving their real-time capability. Four of them were used in the ATAC aboard *Galileo*, making it a 16-bit computer. The ATAC would run five times faster than the attitude-control computer on Voyager.

There were some real pluses in working with this machine. It had floating-point arithmetic capability. It was very flexible. Its 16 CPU registers were not pre-ordained as "accumulator," "address register," or any other specialized component. They could be used for *any* purpose, making it much easier to run multiprocessing software. Also, the user could add instructions to the ATAC's basic set of codes. JPL took advantage of this feature; its software engineers designed four instructions that reduced the system's eventual code size by 1,500 words.

The dual-redundant ATAC computers with 64K memories would be housed in the spun section of the probe, in the ACE (Attitude Control Electronics) boxes. Other parts would be in the DEUCE (de-spun section electronics; they just had to have that acronym!). The DEUCE would be in the lower part of the spacecraft. In a bizarre design decision, the star tracker (used to help maintain spacecraft attitude fixes) was mounted in the spun section! It could only "see" the locking star once every 20 seconds. This was something like trying to keep watch on a lighthouse in the distance, while riding a spinning merry-go-round on a moving flatbed truck!

The CDS could access the memory of the attitude-control computer directly, leaving commands in special locations for the system to "discover" as it ran. The software was written almost entirely in HAL, and ran under a specially-built operating system called GRACOS (Galileo Real-time Attitude Control Operating System). GRACOS scheduled a routine called STARTUP, which in turn scheduled everthing else. Up to 17 processes could run at one time under GRACOS.

The Computer That Was Too Small

There is an incredible irony surrounding spacecraft computers: they need to be small, light, and use little power, but these priorities make failures more likely due to the harsh environment of space.

One fact of life in space flight is that probes have to navigate in areas where electromagnetic and particle fields are dense. With its powerful gravity and plentiful ion production, the Jovian system turned out to be especially dangerous for space probes. When the *Galileo* design decisions were in progress, Voyager had yet to reach Jupiter. When it did, funny things happened. The spacecraft clocks slowed and lost synchronization.

Scientists looked again at some early-1970s Pioneer data. On those missions, some commands had been garbled. At first, everyone was going to blame transient hardware failures.

Eventually, the scientists realized that sulphur ions, ejected from the volcanoes on Io, were being whipped into a high-energy particle field. Particles from this field that struck transistors in the chips of the electronics could flip the bits, causing errors.

In 1982, B. Gentry Lee, the *Galileo* chief engineer (and sometime writing partner of Arthur C. Clarke), studied the problem by testing some of the 2901 processors destined for the attitude-control system. He used a cyclotron in Berkeley to accelerate particles, and let these strike the 2901s. The chips turned out sensitive to what was being called "single-event upsets," or SEUs. SEUs caused data losses in the registers and memories, and corrupted the software.

There was no room in the system for error-correcting codes. The range of possible solutions narrowed: either replace the ATAC with a

continues

continued

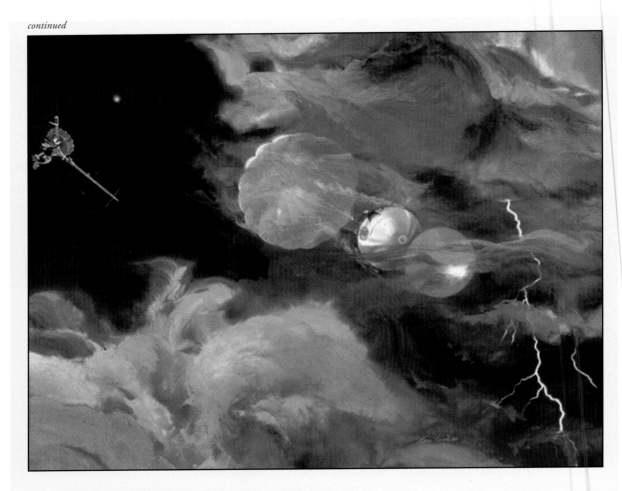

The Galileo probe enters the Jovian atmosphere.

radiation-hardened computer (and take a delay in software development), or have someone rebuild the 2901 architecture in a larger chip, with more empty space between transistors. Lee chose the latter solution. Since the RCA 1802s in the CDS were of an older, less-dense configuration, they did not need to be replaced.

The final irony in this situation is that if the launch had occurred in 1982—on the original schedule—there is a good chance the attitude-control system would have failed just at the critical encounter phase.

Galileo's Tale of Woe

As *Galileo* speeds toward its rendezvous with the gas giant Jupiter, it is struggling with a serious problem. The big high-gain antenna, made of wafer-thin material attached to rods like an umbrella, has failed to deploy completely. Speculation runs high regarding why—from frozen actuators to a design fault caused by having to mount the sun shield just below the antenna on the spun section. When *Galileo* has neared the Sun, ground controllers have tried to point the spacecraft in such a way as to put light and heat on the possibly-frozen motors. The motors have been short-stepped, and other sorts of vibration tried. There is no way to inspect the parts visually, so there is probably no way to figure out exactly what is wrong. Now that the spacecraft is moving away from the Sun, the alternating heat-and-cold treatments are no longer possible.

The main impact of this failure to fully deploy the antenna is a severe reduction in the data rate of the probe's transmissions. (By early 1994, the rate would drop to 10 bits per second—the old Mariner rate.) Since the spacecraft can store some data in the DBUMs and some on tape recorders, it is possible to use these features to buffer transmissions. There is, however, no way the information sent back by *Galileo* will meet the expected volume. Fortunately, *Galileo's* early imaging—of Earth, and an asteroid the probe met on the way to Jupiter—are of the same fidelity as Voyager's. The images that do come back may be fewer, but they should be striking and clear. The key word for *Galileo's* mission is "quality," not "quantity." The versatility of the probe's computers has helped.

Galileo returned this image of the asteroid Gaspra, encountered on its way to Jupiter.

NASA's Legacy and Computers in Future Space Flight

As we near the middle 1990s, we can take stock of NASA's impact on the early history of commercial computing, and also look ahead to the immediate applications of computers in new missions.

The processing demands of NASA's early space missions forced the computer industry into new areas. Embedded computing was completely unknown, and real-time processing (which now exists in such mundane applications as supermarket checkout systems) was limited to the military.

Andromeda galaxy. The sheer scale of space challenges our computing power.

Synergy in the Early Days

NASA began—almost the day it was founded in 1958—to push the state of the computing art in many ways: new uses (simulations, flight control), new components (integrated circuits), new system architectures (networking, redundancy). The pressure of the politically-driven 1960s carried this momentum until the moment of the moon landing. After that, NASA's role as a big driver of computer technology diminished with its budget and influence.

The 1960s marked the era of great synergy between the needs of NASA and the expansion of computing into all walks of life. NASA needed lightweight, low-power, reliable components, so the agency sparked chip production and the design of new integrated circuits. NASA needed high-speed processing of large amounts of data, so the industry (notably IBM) responded with families of mainframes—and sold NASA the high-end models in each case. NASA needed ultra-reliable software, so research labs like Draper had to adopt new verification and validation methods; these are used to test software in embedded systems to this day.

This synergy generated spin-offs readily apparent in the 1990s. They include: safe, computer-based flight-control systems, embedded systems in automobiles, stereos, and nearly every other product that can be helped by a processor, and worldwide networks forming the "Information

Mid-1993 conception of the space station *Freedom* (other versions are sure to come). The two white-banded escape craft are Russian Soyuz spacecraft, still used to service their space station *Mir*.

Launch control at Kennedy Space Center is likely to look like this for many years to come.

Superhighway." All these uses were pioneered in the 1950s and 1960s, when NASA was one of the main players making demands on computer science and software engineering.

Wringing Out the Technology

In the 1970s, computing began to generate its own innate energy for new developments, rather than depending on the needs of special interests. So many organizations, individuals, and industries had come to depend on computing—so thoroughly—that NASA found itself more of a passenger as computer technology took off in the 1970s and 1980s. NASA was able to benefit from this reckless ride; the agency could now choose less expensive, off-the-shelf systems for many of its

needs, rather than paying the entire cost of research and development.

The apparent lag between NASA's use of older computer technology and the "state of the art" is something of an economical example that many computer users might do well to imitate. The tendency in the PC era is to throw out the old computer as soon as the new one is announced—but how many applications that are working satisfactorily really *need* new hardware? There is a point where a machine can no longer work well with its fellows in a network, and is best retired, but routine trashing of two-year-old hardware is a waste of resources that NASA does not do.

The computer industry has more worlds to conquer than NASA alone can provide. Nowadays the computer itself is a tool for exploration of these

The *Cassini* spacecraft and its Titan probe, near Saturn in the early twenty-first century.

Near-Future Uses of Computers in Space

The story of the use of computers in space flight does not end with *Galileo* and the Shuttle. Soon new manned and unmanned missions will need computer power to accomplish their objectives. Let's look at some planned for the 1990s.

Data Processing in Permanent Orbit

President Ronald Reagan announced in his first term that the next big project for the NASA was the construction of the space station *Freedom*. The following decade of design, redesign, budget pressures, and dealing with multiple international partners resulted in no flight hardware and a severely curtailed program.

As this book goes to press, the Japanese and Europeans are still in on *Freedom*, and the Russians are now a major partner—a situation unthinkable in the Reagan years, with Cold War competition still fierce. Plans are to fly nearly 60 astronauts (almost the entire corps) to the Russian space station *Mir* (the name means *Peace*) over the next few years—in preparation for working together on *Freedom*. It is almost impossible to

non-astronomical worlds: artificial intelligence, virtual reality, cyberspace—all artifacts of the Computer Age still unknown to many and explored by relatively few. These are *artificial* worlds—yet, in many ways, as harsh as the environment of space. Space flight, with all its complexity and special conditions, still consists of journeys to natural worlds—which are understandable even if they are severe. Since a computer is an artifact of technology (by definition, not natural), it can be more difficult to understand and integrate into human life than distant planets.

One of the most famous amateur photographs of all time: Aldrin on the moon, with Armstrong reflected in his visor. What was then unthinkable is now unanswerable: Will human beings ever walk the lunar surface again, or will it be the exclusive province of robots?

190

believe that the two space stations, named *Peace* and *Freedom*, may actually reflect the developing political climate. There will be no more Space Races fought like wars; if the public can support *Peace* and *Freedom*, a new era of computing in space will begin.

The space station will have higher data-processing demands than any previous space vehicle. The plan is still to use a network of off-the-shelf processors. IBM has created a prototype—a PC-based network using the AIX operating system—for this purpose. Much of the prototype flight code is in the Ada programming language. There will also be a large ground system to assist in processing and analysis. One goal is to make each workstation in *Freedom* able to display data and use software stored in any other workstation. That way, no matter where the astronauts may be in the station, they have a familiar interface and access to their own data.

The recent addition of the Russians to the partnering arrangement may cause changes in the data-processing system. *Mir* has a reasonably sophisticated set of on-board computers. Although the Russians' use of computers in space once lagged (just as their use of computers in aircraft had done), they have rapidly caught up.

New Simulators for New Missions

Training for space flight will continue to improve in realism. Some of the more sophisticated simulators at Wright-Patterson Air Force Base no longer have analog panels that duplicate the aircraft's controls. Using such panels means representing only one aircraft type at a time. Switching from an F-16 to an F-15 means switching simulators. Now there are large flat screens in the simulators, and the instrumentation on them is displayed using digital graphics. In this way, switching between aircraft means loading a different display program for the cockpit, and a different handling program for any moving base. A single simulator can thus represent a wide variety of aircraft. As airplanes become more computer-intensive, their characteristics will be even easier to duplicate by this method.

Such systems do have limitations. They still need displays and processing for views of the outside world. They do not represent all instruments equally well. The eventual solution lies in *virtual reality*: head-mounted display devices and gloves wired to the system will make it possible for the trainee to see all the instruments and outside views in a single display, and to manipulate them. The large, expensive, and maintenance-hungry motion-base cockpits can then be replaced with a single chair on a ball mount. Other crew can either be simulated, or can participate directly, using their own VR equipment.

Unfortunately, there may not be much to simulate; harsh political and economic weather has all but grounded new aerospace planes on the drawing board. The Japanese and European Space Agencies announced plans for shuttle-like orbiters (much smaller than the U.S. version) a few years ago, and originally slated them to fly late in this decade. Presently the Japanese are backing off from their plans, and the Europeans are extending theirs into the next century. The Russians have grounded their version. The U.S. Space Transportation System is expected to fly until at least 2010, and maybe as late as 2030. Whether it is to be the only one of its kind remains to be seen.

NASA will probably continue to refine the Shuttle, a little at a time. There are plans to upgrade the orbiters' cockpits, replacing many of the gauges and analog instruments with "glass" (digital) displays (as in commercial transport cockpits today). So far, however, there is no clear candidate to succeed the Shuttle as we now know it.

Ground Control: The Present Is the Future

Ground systems will continue to look more like those that do commercial data processing—an ironic role reversal. When the space program began, its ground-control computers were more advanced and powerful than any private installation. NASA was among the first to use distributed processing and networks extensively; now everyone uses them. With the increasing power of hardware and software available to the public, it seems the computer techniques of mission control will not need much innovation. Even so, perhaps some off-the-shelf hardware and software of the 1990s may offer (relatively) cheap solutions to some problems of space flight.

Robot Probes:
The State of the Art

Though the advancements of the PC era are readily available, unmanned space flight also faces difficulties in finding support for new designs. When the Jet Propulsion Laboratory proposed (in the 1980s) a standard data bus for unmanned deep-space probes—to be called "Mariner Mark II"—it was to carry commercial microprocessors and be programmed in the C language. The variety of missions envisioned at the time has been reduced—again. Now the *Cassini* probe to Saturn is the only big science mission in the works. Though it will use some Mariner Mark II concepts, a problem remains: when a mission only

comes along once a decade, standardizing the computer design will be difficult. By the time the next mission is launched, computer technology will have advanced far beyond what its design specified—and retrofitting may be too expensive. Even with NASA's penchant for using proven technology, the rate of change in computers makes using the older technology seem a less intelligent decision. But they may have no choice.

The Past Meets the Future

As *Atlantis* and its crew visits the space station in the twenty-first century, it will still be controlled by five forty-year-old accounting machines. The jump in technology to the computers aboard the Shuttle's successors ought to be breathtaking (at least for historians of computing, if not for the crews). On Earth, we will fly in our computer-intensive airplanes, play our virtual-reality games, and cruise the information superhighway—all partly because NASA wanted its astronauts to return home safely, and its robot probes to get along without constant supervision.

Index

About the Author

Dr. James E. Tomayko is a Senior Member of the Technical Staff of the Software Engineering Institute, and the Associate Director of the Master of Software Engineering Program at Carnegie-Mellon University in Pittsburgh, Pennsylvania.

Dr. Tomayko specializes in embedded computer systems and managing software development. He had a three-year contract from NASA to write a history and evaluation of spacecraft computer systems and ground computers used in support of space flight. He also had contracts from Boeing to work on fault-tolerant digital avionics. Recently he managed the software development for part of a robot that is intended to maintain the Space Shuttle's thermal-protection system.

A licensed pilot, Dr. Tomayko is writing a book on digital flight-control systems. He has written several papers on the history and technology of spacecraft computers for the *Annals of the History of Computing*, *the Journal of the British Interplanetary Society*, and *Aerospace Historian*. His work for NASA was published as a contractor report within the agency, and as an entire volume of the *Encyclopedia of Computer Science and Engineering*.